MATHEMATICAL STUDIES
A Series for Teachers and Students

EDITED BY
DAVID WHEELER
School of Education, University of Leicester

No. 5
CONCEPTUAL MODELS IN MATHEMATICS

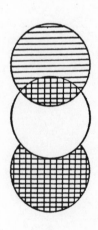

Conceptual Models in Mathematics

Sets, Logic and Probability

by
K. E. HIRST, B.Sc., Ph.D.
Lecturer in Mathematics
University of Southampton

F. RHODES, B.Sc., Ph.D.
Senior Lecturer in Mathematics
University of Southampton

London
GEORGE ALLEN AND UNWIN LTD
RUSKIN HOUSE   MUSEUM STREET

FIRST PUBLISHED IN 1971

© *George Allen & Unwin Ltd* 1971

SBN 004 510034 9 *cased*

SBN 004 510035 7 *paper*

PRINTED IN GREAT BRITAIN
IN 10 ON 12 PT TIMES
BY PAGE BROS (NORWICH) LTD
NORWICH

FOREWORD

The developments in mathematics over the last hundred years have brought a different understanding of the nature of mathematical activity which has inevitably, though slowly, led to reappraisals of the mathematics that should be taught and appropriate ways of teaching it. It is no longer possible to believe that developments in mathematics concern only the research mathematician and do not have any bearing on the mathematics taught in schools. This series of books is intended as a contribution to the reform of school mathematics by introducing to the reader some areas of mathematics which, broadly speaking, can be called modern, and which are influencing the content of school syllabuses.

The series does not put forward explicit advice about what mathematics to teach and how it should be taught. It is meant to be useful to those teachers and students in training who want to know more mathematics so that they can begin to take part in teaching the new courses, or modifying them, or devising their own syllabus revisions, however modest. The books are elementary without being trivial: the mathematical knowledge they assume is roughly that of a traditional grammar school course, although substantial sections of all the books can be understood with less.

Now that the stability over a long period of school mathematics syllabuses has come to an end, it is to be hoped that a new orthodoxy does not suceed the old. The reform of mathematics teaching should be a continuing process, associated with a deepening study of the subject throughout every teacher's professional life. These books may help to start some teachers on that course of study.

D.W.

PREFACE

This book arose from a course given for first-year mathematics students at the University of Southampton, and a preliminary version has been used there for three years. It was also discussed in a seminar course for High School mathematics teachers at Wesleyan University, Connecticut, while the second author was a visiting Professor there. In the book we have tried to capture some of the action of our classroom discussions. We concentrate on two mathematical activities. The first is the creation of mathematical models. The second is the communication of mathematical proofs.

It is a pleasure to express our thanks to our colleagues and students. We have learned much from them about mathematics and about the ways in which people progress in their understanding of mathematics. Many points raised by them have been incorporated into this book. In particular, we are indebted to Dr F. A. Bostock for discussions of the material in the first three chapters, and to Mr T. M. F. Smith for suggestions concerning the chapter on probability. Others whose comments have been most helpful include Prof. Ethan Coven, Mrs Louise Rosenbaum, Mr C. L. Thompson and Dr A. G. Howson.

We are grateful to the Mathematics Department of Southampton University for making the use of the preliminary version possible, and to the publisher for allowing us to continue to use this version whilst the book was in the press.

K.E.H. and F.R.

CONTENTS

CONTENTS

LIST OF SYMBOLS

Symbol	Description	Page
$\in$	belongs to; is a member of	17
N	set of natural numbers	18
Z	set of integers	18
Q	set of rational numbers	18
R	set of real numbers	18
$\{a, b, c\}$	set described as a list	18
$\{x \mid P(x)\}$	set described by a rule	19
$A \cup B$	union of A and B	25
$A \cap B$	intersection of A and B	26
A^c	complement of A	28
$A \backslash B$	relative complement of B with respect to A	28
$A \triangle B$	symmetric difference of A and B	29
$\varnothing$	empty set	30
$A \subseteq B$	A is a subset of B	31
$A \subset B$	A is a proper subset of B	32
$\mathscr{P}(A)$	power set of A	33
$\sim$	negation	43
$\wedge$	and	44
$\vee$	or	44
$\Leftrightarrow$	if and only if	45
$\Rightarrow$	implies	45
$\forall$	universal quantifier	56
$\exists$	existential quantifier	56
(x, y)	ordered pair	77
$A \times B$	cartesian product of A and B	78
$\mathscr{D}(S)$	domain of S	80
$\mathscr{R}(S)$	range of S	80
S^{-1}	inverse relation to S	83
$T \circ S$	composition of T and S	83
I_A	identity relation on A	84
A/S	quotient set	89

Symbol	Description	Page
$f: A \rightsquigarrow B$	f is a function from A to B	99
$*$	binary operation	114
$(G, *)$	group	122
(X, ρ)	metric space	132
$(U, \mathscr{A}, m)$	measure space	141
(S, P)	finite probability space	150
$P(F \mid E)$	conditional probability of F given E	157

Notes

1. There is a variety of notations used for the complement of a set A. In place of the notation A^c, other authors use $\mathscr{C}A$ or A' or $\overline{A}$.

2. In place of the notation $A \backslash B$ for relative complement, other authors use $A - B$ or $A \frown B$.

3. We have chosen the notations $\subseteq$ and $\subset$ for use with subsets and proper subsets because of the analogy with the notations $\leqslant$ and $<$. Some authors adopt the notation $\subset$ for use with subsets and have no separate notation for use with proper subsets.

4. Many authors use $f : A \rightarrow B$ as a notation for the statement 'f is a function from A to B'. The arrow $\rightarrow$ is used in analysis in connection with limiting processes, and so we have used the arrow $\rightsquigarrow$ in the notation $f : A \rightsquigarrow B$.

INTRODUCTION: MATHEMATICAL MODELS

Many phenomena which are observed in everyday life are of a very complex nature. Men have tried to understand them by comparing a number of similar instances of a phenomenon. They have focused on a few aspects of interest to them and have tried to see how they are related. Deepening understanding of a phenomenon has usually been linked with the development of a language and notation through which this understanding can be expressed. Thus an increasing understanding of the layout of towns and buildings was accomplished by and through the development of Euclidean geometry. Situations which include changing velocities were clarified when the language of calculus was devised, and when the concept of instantaneous velocity was used in conjunction with it.

These branches of mathematics serve as conceptual models of certain aspects of the physical world. For some purposes one needs very general models which can be applied widely even though they cover few aspects of any particular situation to which they are applied. For other purposes one needs specialized models which cover more aspects of a particular situation, but which are less widely applicable. Most phenomena can be studied through several models depending upon which aspects one is interested in. In choosing an established model through which to study a particular situation one tacitly accepts the premises on which the model was built. These will usually include some conventions concerning the way in which the model is to be fitted to a particular situation.

In many cases the drive to understand phenomena has come from a desire to control them or at least to make predictions about them. Thus two considerations need to be kept in mind when one tries to develop or choose a conceptual model through which to study a particular situation. Firstly the model should be easy to work with, and secondly it should take into account all significant characteristics and fit them well. These two criteria may well act in opposition. For example, Euclidean geometry is easy to work with, and for use in small-scale surveys on flat ground the identification of the points and lines of geometry with points and lines on the ground is subject to the relatively minor restrictions imposed by undulations of the ground and by the limitations of survey instruments. However, in a survey of a continent, mountain ranges necessitate greater approximations. Moreover, the curvature of the earth is a significant factor, and spherical geometry may be used to obtain a better fit, even though it is harder to work with than Euclidean geometry.

The two criteria appear not only in the choice of an established model for a particular situation, but also in the creation of new models. Some aspects of a model will be devised to fit the characteristics under consideration. Others will be devised to make the model easy to work with. We shall illustrate the use of these two criteria in the models discussed in this book.

The most general model which we consider is the language of sets. Nearly all of its aspects arise from the study of everyday language concerning collections of objects and of the use of notation in mathematics. It can be applied to any situation in which one is concerned with putting collections together and splitting them up. In addition to its breadth of application it has a second great value. It is of such generality that it can be used as the foundation for many other mathematical theories. In particular it can be used as the basis for a definition of the natural numbers, and also of a collection of transfinite numbers whose investigation has yielded many of the most important twentieth-century contributions to mathematics.

The models which are introduced in the last three chapters of the book are more specialized and so are of more limited application. Relations, functions and binary operations are treated in such a way as to reveal the relationship between the concepts and the situations which they model. Metric spaces, measure spaces and probability spaces serve as other examples of conceptual models which are based on the theory of sets. The internal structures of the various concepts are carefully studied, and the topics taken far enough to show the power of their use in mathematics. For example, we include the factorization of a function into an injection and surjection via a quotient set of the domain. We also include the definition of quotient groups of abelian groups. We have not assumed a background of calculus for this book. However, for those readers who have some knowledge of calculus, we include in the section on metric spaces a discussion of spaces of polynomials.

The models we have chosen to study are included because of their basic importance in mathematics, and because they provide increasingly difficult examples of proofs of statements within the models. Proofs as they are commonly presented by mathematicians consist of somewhat stylized sentences which serve to guide other mathematicians through a sequence of logical steps. In order to show the connection between these two aspects we devote Chapter 2 to a discussion of truth tables, tautologies, quantified statements and mathematical proof. We present many proofs in two styles, the discursive style normally used by mathematicians and a formal style which exposes the logical basis of the proof.

Our discussions of sets and proof are designed to create models which are sufficiently detailed for use in later parts of the book, while at the same time to reveal the genesis of the models. For other purposes more formal presentations of set theory and logic are necessary. As an introduction to more formal theories we recommend P. R. Halmos, *Naive Set Theory* (van Nostrand) and R. R. Stoll, *Sets, Logic and Axiomatic Theories* (W. H. Freeman).

1

THE LANGUAGE OF SETS

1.1 INTRODUCTION

When we make use of collections of objects in various ways, we can distinguish two situations: those where order is important and those where it is not. These two situations can best be considered by means of examples, and we give examples from within and from outside mathematics.

Firstly, here are some situations in which order is important. The order in which letters are used in the spelling of words is important. For example, the word 'but' is different from the word 'tub', although both words use the same letters. Similarly, the order in which digits are used to express numbers in the scale of ten is important. For example, 342 is a greater number than 324. When points in space are described relative to a cartesian system of axes, the points $(12, 4, 35)$ and $(35, 4, 12)$ are different from each other, so once again, order is important.

As an example of a situation in which order is not important, we might consider a description of the even numbers between 1 and 9. It then makes no difference whether we write $2, 4, 6, 8$, or $8, 6, 4, 2$, or indeed any other ordering of the four numbers. The information we wish to communicate will be imparted by any of these descriptions. Similarly, when we wish to describe a triangle in the plane by listing its vertices, then the order in which we list the vertices is immaterial for that purpose.

Now for a non-mathematical example, we can consider a class of students. If we wish to know which students are in the class, then the order in which their names appear on a list is immaterial. It may be convenient to have a list of names in alphabetical order, but that will not be essential. On the other hand, if we publish the relative standing of students after an examination without publishing the examination marks themselves, then the order in which names appear on the list is vital.

We shall discuss both types of situation in this book. In the remain-

15

der of this chapter we shall look at the notion of sets which arises from the consideration of situations in which order is not important. In Chapter 3 we shall base the study of situations in which order is important on the notion of ordered pairs.

1.2   SETS

In the introduction, the examples of situations in which the order of description is not important were relatively simple. In each case the objects of a given collection were all of the same kind, and there would be no difficulty in deciding what were the objects in the collection. Thus in a class of students, the 'objects' are all people, and it is easy to decide which people are in a class. However, not all collections of objects are as simple as that. If you empty your pockets you will obtain a collection with a great variety of kinds of objects—a handkerchief, a key-case, a wallet, a diary, a pen, and so on. Not only are the objects of widely differing kinds, but also you may have difficulty in deciding what should be regarded as an 'object'. Should each key be regarded as an object separate from the key-case? Should each piece of paper in the wallet be regarded as an object separate from the wallet? Should each of the pieces of leather which make up the wallet be regarded as a distinct object in the collection of objects in your pockets? There is no unique correct answer to any of these questions. At any time you will describe the contents of your pockets to the extent that the occasion warrants, and you will bear in mind that some things may be regarded as single objects in a larger collection or they may be regarded as collections of their component parts. Similarly, in a list of the contents of your house, you might list as single objects a tea service and a chess set, but when you come to use either of them you will regard each as a collection and make sure that each of its component objects is there.

The mathematical language of set theory is a conceptual model of our understanding of the way that collections of physical objects, described with no regard to order, can be combined in various ways. Where the manipulation of collections of physical objects gives enough guidance, this is used to develop the language of set theory. But where a question arises in the development of the language of sets for which collections of physical objects give no clear guidance as to an answer, the language is developed in such a way as to make it as easy to handle as possible and as free from exceptional cases as possible.

The basic notation for set theory is of the form

$$a \in A$$

which is read 'a belongs to A'. The negation of this statement, 'a does not belong to A', is denoted by $a \notin A$. It is required that for every pair of symbols a, A which are in use at any time, precisely one of the statements $a \in A$ and $a \notin A$ is true. With respect to the statement $a \in A$, the symbol a plays the role of an element and the symbol A plays the role of a set. However, just as in the physical world a tea service is regarded sometimes as a single object and sometimes as a collection of objects, in set theory there is no clear-cut division between elements and sets. The symbol A, which plays the role of a set in the statement $a \in A$, plays the role of an element in the statement $A \in \mathscr{A}$. Similarly, the symbol a plays the role of a set in the statement $\alpha \in a$. Nevertheless, we shall use the terms *set* and *element*, just as we use the terms 'collection' and 'object', in any context in which we can do so without undue ambiguity.

The equality sign is used in set theory in a way which models the notion that two collections of objects are the same. In everyday language phrases which express equality of collections of objects are used in several different senses. For example, if Tom and Joe are said to have the same friends, then it is understood that every boy is a friend of both or of neither. On the other hand, if Tom and Joe are said to use the same textbooks it will probably mean that if one uses a particular textbook then the other uses another copy of the same textbook, although it might mean that they have only one set of books between them. The ambiguities involved in the use of 'the same' in these expressions are similar to the ambiguities involved in everyday expressions about set membership. However, in each case the implication is that two collections of objects are equal if whenever an object belongs to one collection then 'the same' object belongs to the other. In view of this, the following rule is chosen for the use of the equality sign in set theory. The symbol $A = B$ is used in the following case and in that case only: if $a \in A$ then $a \in B$, and also if $a \in B$ then $a \in A$. In other words, $A = B$ is used as an abbreviation for '$a \in A$ if and only if $a \in B$'. This can be described by the statement that two sets are equal if and only if they have the same elements.

We shall be interested in applications of the language of sets to other branches of mathematics. In these applications a set is regarded as a collection of mathematical concepts or symbols specified in such a way that it is known for each concept and symbol whether or not it belongs to the set. For the applications it is necessary to have notations for collections of concepts such as numbers and functions, and for collections of mixed symbols, described with no regard to order.

Some sets of numbers are used so frequently that special letters are assigned to them by authors for use throughout their books. However, authors do not necessarily agree with one another about the assignment of letters to these sets. We shall use the symbol N to denote the *natural numbers* (i.e. the numbers $1, 2, 3$, etc., which alternatively will be called the positive integers), Z to denote the *integers* (i.e. the set N together with zero and the negative whole numbers), Q to denote the *rational numbers* (i.e. numbers which can be expressed in the form a/b where a and b are integers with $b \neq 0$), and R to denote the *real numbers*.

The notation commonly used for other collections of mathematical objects consists of a concise description of the collection placed between braces { }. If the collection contains only a few objects, it may be convenient to list them all, as in the following examples.

$$\{4, 5, 6, 7, 8\},$$
$$\{\cos, \sin, \exp, \log\},$$
$$\{\Delta, 54, \#, \lambda, a, abc, \{a, b, c\}, cab\}.$$

While we regard the first as a set of numbers and the second as a set of functions, in the third example a miscellaneous collection of symbols has been used. In this notation for sets punctuation marks cannot themselves be used as symbols, and braces are used in pairs in such a way that an expression enclosed in braces is separated from other symbols by commas. With respect to this notation for a set, the elements are denoted by the symbols cut off by the commas (the commas within subsidiary braces being ignored). Thus

$$4 \in \{4, 5, 6, 7, 8\},$$

and the number 4 at the beginning of this statement can be replaced by 5, 6, 7 and 8, but by no other number. Also

$$\sin \in \{\cos, \sin, \exp, \log\},$$

and the function sin at the beginning of this statement can be replaced by cos, exp and log, but by no other function. Similarly

$$a \in \{\Delta, 54, \#, \lambda, a, abc, \{a, b, c\}, cab\},$$

and the symbol a at the beginning of this statement can be replaced by $\Delta, 54, \#, \lambda, abc, cab$ and by $\{a, b, c\}$. Note that in the statement

$$\{a, b, c\} \in \{\Delta, 54, \#, \lambda, a, abc, \{a, b, c\}, cab\},$$

the symbol $\{a, b, c\}$ is playing the role of an element, whereas in the statement $a \in \{a, b, c\}$ it is playing the role of a set.

If a collection of mathematical objects is too large for a full list to be given as above, then it may be described by means of properties which are satisfied by the objects in the collection and by no others. Use may be made of such familiar ideas as algebraic variables and inequalities like $3 < x < 9$. The collection of real numbers between 3 and 9 can be thought of as the collection of objects x such that x is a real number and $3 < x < 9$. With the standard notation for the set of real numbers introduced above, this is the collection of objects x such that $x \in R$ and $3 < x < 9$. The corresponding notation used in set theory is

$$\{x \mid x \in R \text{ and } 3 < x < 9\}.$$

With respect to this notation for a set, the elements are the specific mathematical objects which satisfy the rule following the vertical line. Thus, for example,

$$5\tfrac{1}{2} \in \{x \mid x \in R \text{ and } 3 < x < 9\},$$
$$2\tfrac{1}{2} \notin \{x \mid x \in R \text{ and } 3 < x < 9\}.$$

The two methods of denoting sets may be called, for convenience, the list method and the rule method. For sets which have been described by the list method, the question of equality of sets can be resolved by direct reference to the rule for the use of the equality sign in set theory. Consider the sets

$$\{4, 5, 6, 7, 8\}, \{5, 4, 8, 7, 6\}, \{5, 5, 4, 7, 8, 6, 7\}.$$

The first two are equal, as can be seen by checking that $4 \in \{4, 5, 6, 7, 8\}$ and $4 \in \{5, 4, 8, 7, 6\}$, and similarly for 5, 6, 7 and 8, while no other element belongs to either set. The third set may have arisen if a collection of numbers had been formed by seven people separately choosing a digit. Now according to the rule for the use of the symbol $\in$ in relation to this notation for sets,

$$5 \in \{5, 5, 4, 7, 8, 6, 7\},$$

and the number 5 at the beginning of this statement can be replaced by 4, 6, 7 and 8, and by no other number. Thus the repetition of 5 and 7 within the braces is of no significance, and we have

$$\{5, 5, 4, 7, 8, 6, 7\} = \{4, 5, 6, 7, 8\}.$$

On the other hand

$$\{4, 5, 6, 7, 8\} \neq \{4, 5, 6, 7\},$$

since $8 \in \{4, 5, 6, 7, 8\}$ whilst $8 \notin \{4, 5, 6, 7\}$.

To determine whether two sets which have been presented by the rule method are equal or not, it is necessary to appeal to that part of mathematics in which the rules are expressed. Thus, by reference to our knowledge of numbers, and in particular to the fact that $5\frac{1}{2}$ is a real number but not an integer, we can determine that

$$\{x \mid x \in R \text{ and } 3 < x < 9\} \neq \{x \mid x \in N \text{ and } 3 < x < 9\},$$

since $5\frac{1}{2} \in \{x \mid x \in R \text{ and } 3 < x < 9\}$, whilst $5\frac{1}{2} \notin \{x \mid x \in N \text{ and } 3 < x < 9\}$. On the other hand, consider the two sets

$$A = \{x \mid x \in R \text{ and } -2 < x < 2\} \text{ and } B = \{x \mid x \in R \text{ and } x^2 < 4\}.$$

If $a \in A$, then $a \in R$ and $-2 < a < 2$, so that $a^2 < 4$, from which it follows that $a \in B$. Moreover, if $b \in B$, then $b \in R$ and $b^2 < 4$, so that $-2 < b < 2$, from which it follows that $b \in A$. We have thus shown that

$$\{x \mid x \in R \text{ and } -2 < x < 2\} = \{x \mid x \in R \text{ and } x^2 < 4\}.$$

Similar reference to properties of the integers will show that

$$\{x \mid x \in Z \text{ and } 3 < x < 9\} = \{x \mid x \in N \text{ and } 3 < x < 9\}$$
$$= \{x \mid x \in N \text{ and } 4 \leqslant x \leqslant 8\} = \{4, 5, 6, 7, 8\}.$$

Now the list method of presenting sets can be extended to sets with large numbers of elements if some of the terms are omitted and replaced by three dots. Thus the set $\{x \mid x \in Z$ and $3 < x < 29\}$ can be written as $\{4, 5, 6, \ldots, 28\}$. With this form of presentation it is necessary to include enough elements before the dots to indicate the form of the progression. If the progression terminates, as in the above example, the final element is also included. However, if the progression does not terminate, the expression within the braces ends with the dots. For example, an alternative way of denoting the set $\{3n \mid n \in N\}$ is $\{3, 6, 9, 12, \ldots\}$.

Upon meeting the list method of presenting sets for the first time one must think carefully about the significance of the order in which the elements occur. If no dots are used in the list then the order of the list can be altered at will. But if dots are used, the order of the first few elements is needed to indicate the form of the progression. However, it can be argued that such a list does not define a set at all, since the person who reads the list is forced to put some interpretation on it and this may not coincide with that of the person who wrote it down. Thus the statement

$$S = \{3, 5, 7, 11, 13, \ldots\}$$

admits several possible interpretations. The reader could decide either that S is the set of all odd primes, or that S is the set of all odd natural numbers which are not perfect squares. This could perhaps be countered by the claim that we have not listed enough elements; but even if we do list more, similar ambiguities may occur. Why then is the list method with dots used at all? It is used on the basis that there are some patterns which are so common a part of our mathematical experience that there is very little danger of misinterpretation. Thus if we write

$$A = \{1, 4, 9, 16, 25, \ldots\},$$
$$B = \{2, 4, 6, 8, 10, \ldots\},$$
$$C = \{2, 4, 8, 16, 32, \ldots\},$$

then we are almost certain that the reader will interpret A as the set of all perfect squares, B as the set of all even natural numbers, and C as the set of all powers of 2. The context in which such a statement is used will often make the meaning of the list quite clear. However, before you dismiss the problem of the ambiguities which are inherent in this method of describing sets, consider exercise 6 at the end of this section.

A second kind of ambiguity arises with the list method of describing sets for application to a branch of mathematics in which the concepts admit more than one notation. For example, relative to the natural numbers, the symbols 5, V, five, cinq, $4+1$, $27-22$, etc., denote the same concept. What then should be the reaction to

$$\{5, \text{five}, 4+1\}?$$

That it is ambiguous. Out of context, one cannot tell whether the author intended the reader to place the emphasis on the difference between the three symbols or on the uniqueness of the number concept which they represent. This particular example is not likely to arise in any application of sets to the study of numbers. However, the example

$$\left\{\frac{2}{4}, \frac{3}{6}, \frac{1}{3}, \frac{2}{6}\right\}$$

could easily arise in a discussion of the rational numbers, and this suffers from the same ambiguity as the previous example. Since 2/4 and 3/6 are different fractional forms for the rational number 'a half', we have

$$\left\{\frac{2}{4}, \frac{3}{6}, \frac{1}{3}, \frac{2}{6}\right\} = \left\{\frac{1}{2}, \frac{1}{3}\right\},$$

if the symbols stand for rational numbers, while

$$\left\{\frac{2}{4}, \frac{3}{6}, \frac{1}{3}, \frac{2}{6}\right\} \neq \left\{\frac{1}{2}, \frac{1}{3}\right\}$$

if the symbols stand for fractional forms. Whenever any ambiguity of this type might appear in your work, you must make it quite clear what concepts the symbols are meant to represent.

The rule method of presenting sets must also be used with care. Each rule must be expressed in such a way that the author and the readers will agree on which objects belong to the set and which do not. It is easy to avoid rules which are meaningless, as in

$$\{x \mid x \text{ is a fierce number}\},$$

but not so easy to avoid rules which are ambiguous, as in

$$\{x \,|\, 3 < x < 9\}.$$

The latter notation will be taken by some readers to refer to the integers between 3 and 9 and by some to refer to the real numbers between 3 and 9. There will be no ambiguity of this kind however, if a prior decision has been reached to restrict the discussion to the integers, say. In this case the restriction $x \in Z$ applies throughout, and need not be restated. This is particularly useful when several sets have to be defined and relationships between them discussed. When an overall restriction is put on the type of object to be allowed in any piece of mathematics, the collection of all objects of the appropriate type is called the *universe of discourse*, or the *universal set* for that piece of work. Thus there is no ambiguity if one defines three sets as follows: let the universe of discourse be Z and let

$$A = \{x \,|\, 3 < x < 9\}, \quad B = \{x \,|\, x > 27\}, \quad C = \{x \,|\, 3 < x < 29\}.$$

Note that the ambiguity inherent in the notation

$$\left\{\frac{2}{4}, \frac{3}{6}, \frac{1}{3}, \frac{2}{6}\right\},$$

which was discussed earlier, can be resolved if one states either that the universe of discourse is the set Q of rational numbers, or that the universe of discourse is the set of fractional forms. You should avoid ambiguity in each piece of your own work in which you use a universe of discourse by stating it clearly, and you will need to discover in other people's work the universe which they have assumed when they have not stated it explicitly.

It sometimes happens that the symbolic expression of the rule which defines a set is longer than is desirable if it has to be repeated often, and in this case it is convenient to substitute for it a brief symbol. For example if we let $P(x)$ stand for the sentence 'x is an even integer which is not divisible by 3 or by 5' we can write $\{x \,|\, P(x)\}$ for the set $\{x \,|\, x$ is an even integer which is not divisible by 3 or by 5$\}$. Note that each set A can be expressed by the rule method by use of the notation $x \in A$. The equality $A = \{x \,|\, x \in A\}$ will prove to be of particular value in the rest of the chapter.

Exercises 1.2

1. Discuss with other people examples of rules which are sufficiently precise to define sets and examples of rules which do not define sets because membership cannot be determined.

2. State which of the following statements are true.
   (a) $4 \in \{5, 7, 4, 2\}$.
   (b) $7 \in \{4, 5, 3\}$.
   (c) $\{a, \varDelta\} \in \{a, \varDelta, b, \#\}$.
   (d) $\{\lambda\} \in \{a, b, 5, \{\lambda\}, p\}$.
   (e) $5 \in \{x \mid x \in R \text{ and } x > 0\}$.
   (f) $4 \in \{x \mid x \in N \text{ and } x^2 + 2x - 1 = 0\}$.
   (g) $\{4, 5, 6, 7, 8\} = \{7, 4, 8, 5, 6\}$.
   (h) $\{7, a, 3, \varDelta\} = \{7, 7, \varDelta, 3, a, 3\}$.
   (i) $\{a, b, c, d\} = \{a, \{b, c\}, d\}$.
   (j) $\{x \mid x \in R \text{ and } x^2 = 1\} = \{x \mid x \in Q \text{ and } x^2 = 1\}$.
   (k) $\{x \mid x \in N \text{ and } 0 < x \leqslant 5\} = \{0, 1, 2, 3, 4, 5\}$.
   (l) $\{2n \mid n \in N\} = \{2, 4, 6, 8, \ldots\}$.

3. (a) Write the set $\{x \mid x \in N \text{ and } 0 < x < 10\}$ using the list method.
   (b) Write the set $\{2, 4, 8, 16, 32\}$ using the rule method.
   (c) Write the set of real numbers which are solutions of the equation $x^2 - 3x + 2 = 0$, using the rule method and using the list method.
   (d) Write a general expression for a set consisting of 20 elements.

4. Let the universe of discourse be the set N of natural numbers. Write the following sets by the list method.
   (a) $\{x \mid 3x - 4 = 5\}$.
   (b) $\{x \mid x^2 - x - 20 = 0\}$.
   (c) $\{x \mid x^2 - 1 = (x - 1)(x + 1)\}$.

5. Can you write the set $\{x \mid x \in N \text{ and } x^2 - x + 6 = 0\}$ by the list method?

6. For each natural number n, let n points be chosen on the circumference of a circle so that when each pair of points is joined by a line segment no three line segments meet at a point inside the circle. Let $f(n)$ denote the number of regions into which the inside of the circle is divided by the segments. From diagrams, find the values $f(1)$, $f(2)$, $f(3)$ and $f(4)$, and discuss with other people any predictions which these four values lead you to make concerning $f(5)$,

$f(6)$, etc. Calculate $f(5)$ and $f(6)$ from diagrams and discuss your reactions to these.

1.3 OPERATIONS WITH SETS

In the preceding section we considered ways of introducing particular sets. In this section we shall be concerned with the first elementary manipulations of sets, and these will give rise to new ways of describing sets. There are many occasions on which two sets A and B are known and one needs to consider those elements which belong to both A and B, and those elements which belong to one set but not the other, and those elements which belong to at least one of A and B. The language and symbols which have been developed to do this easily are as follows.

Union

The elements which belong to at least one of A and B form a set which is called the *union* of A and B. This set is denoted by $A \cup B$, which may be read as 'A union B'. In the rule method of describing sets,

$$A \cup B = \{x \mid x \in A \text{ or } x \in B\}.$$

Here the word 'or' is used inclusively in the sense of and/or, i.e. the condition defining $A \cup B$ is understood to mean $x \in A$ or $x \in B$ or both.

The expression 'either $x \in A$ or $x \in B$' will be used to mean $x \in A$ or $x \in B$, but not both. This is used on p. 29 to define the symmetric difference of two sets. An explanation of the use of 'or' and 'either-or' in mathematical statements and proofs is given in §2.3.

Specific examples of unions of sets are

$$\{3, 4, 5\} \cup \{4, 5, 6\} = \{3, 4, 5, 6\},$$

$$\{x \mid x \in R \text{ and } x > 3\} \cup \{x \mid x \in R \text{ and } 0 < x < 5\}$$
$$= \{x \mid x \in R \text{ and } x > 0\}.$$

It is convenient to have a pictorial representation of the set operations. To this end we introduce the device known as a Venn diagram.

Venn diagram

Regions of the plane bounded by closed curves are used to represent individual sets, and shading is used to indicate the result of performing a given operation. Venn diagrams will be useful when expressions involving several operations are considered, and when relationships between such expressions are discussed.

A Venn diagram illustrating the union of two sets is shown in

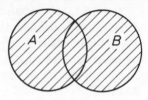

Figure 1.1

Fig. 1.1. The two sets *A* and *B* are represented there by the two circular regions while the shaded region represents the set $A \cup B$.

Intersection

Those elements which belong to both *A* and *B* form a set which is called the *intersection* of *A* and *B*. It is denoted by $A \cap B$, which may be read as '*A* intersection *B*'. Thus '$x \in A \cap B$' is an abbreviation for '$x \in A$ and $x \in B$', and so in the rule method of describing sets

$$A \cap B = \{x \mid x \in A \text{ and } x \in B\}.$$

For example

$$\{3, 4, 5\} \cap \{4, 5, 6\} = \{4, 5\},$$

$$\{x \mid x \in R \text{ and } x > 3\} \cap \{x \mid x \in R \text{ and } 0 < x < 5\}$$

$$= \{x \mid x \in R \text{ and } 3 < x < 5\}.$$

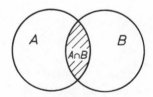

Figure 1.2

A Venn diagram illustrating the intersection of two sets is shown in Fig. 1.2.

Extended notation

The notations for union and intersection can be extended to cases where more than two sets are being considered. If A, B and C denote given sets then the elements which belong to at least one of these three sets form a set

$$A \cup B \cup C = \{x \mid x \in A \text{ or } x \in B \text{ or } x \in C\}$$

called the union of A, B and C. If we change the notation for the three sets to S_1, S_2, S_3, then the rule for defining the union can be written

$$S_1 \cup S_2 \cup S_3 = \{x \mid x \in S_i \text{ for some } i = 1, 2, 3\}$$

where 'for some $i = 1, 2, 3$' is an abbreviation for 'for $i = 1$ or $i = 2$ or $i = 3$'. This can be extended to the case of the union of n sets for which the following notations are used

$$\bigcup_{i=1}^{n} S_i = S_1 \cup S_2 \cup \ldots \cup S_n = \{x \mid x \in S_i \text{ for some } i = 1, 2, \ldots, n\}.$$

The first symbol is read as 'the union of S_i from $i = 1$ to $i = n$' while the second can be read as 'S_1 union S_2 union — to S_n'. Similarly, the set of those elements which belong to A and B and C is called the intersection of these sets. The notations for intersections corresponding to those given above for unions are

$$A \cap B \cap C = \{x \mid x \in A \text{ and } x \in B \text{ and } x \in C\},$$

$$S_1 \cap S_2 \cap S_3 = \{x \mid x \in S_i \text{ for each } i = 1, 2, 3\},$$

$$\bigcap_{i=1}^{n} S_i = S_1 \cap S_2 \cap \ldots \cap S_n = \{x \mid x \in S_i \text{ for each } i = 1, 2, \ldots, n\}.$$

Complement

Recall that in certain pieces of mathematics a universe of discourse for the objects under discussion may be agreed. If a universe of discourse U has been agreed then given any set A we may consider those objects which do not belong to A. They form a set, called the *complement of* A, which we shall denote by A^c. In the rule method of describing sets,

$$A^c = \{x \mid x \notin A\},$$

it being understood that the variable x is restricted to the universe of discourse agreed upon. Thus '$x \in A^c$' is an abbreviation for '$x \in U$ and $x \notin A$'. You will see that if you decide upon Z as the universe of discourse and $\{1, 2, 3\}$ as your set, while someone else decides upon N as the universe of discourse and $\{1, 2, 3\}$ as his set, then the two sets are equal, but their complements are different because his complement consists only of the natural numbers $4, 5, 6, \ldots$ while your complement includes also all the negative integers.

In illustrating the set A^c a larger region is used to represent the universe of discourse U, and A^c is shaded as indicated in Fig. 1.3.

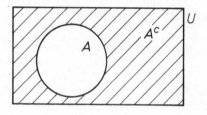

Figure 1.3

Difference

The set of elements which belong to A but not to B is called the *relative complement of* B *with respect to* A, and is denoted by $A \backslash B$. The name is sometimes abbreviated to the difference A minus B. In the rule method,

$$A \backslash B = \{x \mid x \in A \text{ and } x \notin B\}.$$

Similarly, the relative complement of A with respect to B is

$$B \backslash A = \{x \mid x \in B \text{ and } x \notin A\}.$$

Note that if the universe of discourse is U, then $U \backslash A = A^c$.

Symmetric difference

In discussions of sets, the union of $A\backslash B$ and $B\backslash A$ is used sufficiently often for it to be desirable to have a special symbol for this set. The symbol $A \triangle B$ is used. The set $A \triangle B = (A\backslash B) \cup (B\backslash A)$ is called the *symmetric difference of A and B*. It is the set of elements belonging either to A or to B, but not to both. The statement 'either $x \in A$ or $x \in B$' will be used in the sense of 'but not both', and so

$$A \triangle B = \{x \mid \text{either } x \in A \text{ or } x \in B\}.$$

Illustrations for the operations of difference and symmetric difference are given in Figs. 1.4 (a) and 1.4 (b) respectively.

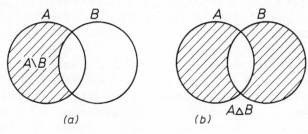

Figure 1.4

The empty set

In the description of the operations given above, it has been assumed that given any universe of discourse U and any pair of sets A and B in U, there are sets $A \cup B$, $A \cap B$, A^c, B^c, $A\backslash B$, $B\backslash A$, and $A \triangle B$ in U. Thus we had the example

$$\{3, 4, 5\} \cap \{4, 5, 6\} = \{4, 5\}.$$

However, one cannot write $\{1, 2\} \cap \{3, 4\}$ in the form of a list, neither can one write $\{1, 2\} \triangle \{1, 2\}$ in the form of a list. Indeed, when A and B have no elements in common,

$$A \cap B = \{x \mid x \in A \text{ and } x \in B\}$$

contains no elements. When $A = U$,

$$A^c = \{x \mid x \in U \text{ and } x \notin A\}$$

contains no elements. When $A = B$,

$$A \triangle B = \{x \mid x \in A \text{ and } x \notin B\} \cup \{x \mid x \in B \text{ and } x \notin A\}$$

contains no elements.

It is not usual to consider collections of physical objects which contain no objects. However, the language of sets leads us to consider general expressions involving intersections, complements and differences, and it would be exceedingly inconvenient to have to regard the circumstances described above as leading to exceptional situations which lie outside the language of sets. We therefore extend the language to include them by saying that $\{1, 2\} \cap \{3, 4\}$ is an empty set, and more generally if A and B have no elements in common then $A \cap B$ is said to be an empty set. We now have to ask how many empty sets we shall need to introduce, and to answer this we must attempt to interpret for empty sets the statement about equality of sets which was made in §1.2. That statement was that two sets are equal if they contain the same elements. Thus two sets, both of which are empty, are equal. If you wish to disprove this point then you must show that the two sets are unequal, and to do this you must show that one of them contains an element which does not belong to the other set — and that you cannot do.

You may well feel that these arguments are not convincing and you will perhaps quarrel with them on linguistic grounds. This would suggest that we need to delineate more carefully the arguments we use in proving statements, and this task will be undertaken in the next chapter. In terms of the formalized language of proof developed there we shall show that there is just one set with no elements, and we call it the *empty set* (or the *null set*) and denote it by $\varnothing$.

Exercises 1.3

1. Let the universe of discourse be $U = \{1, 2, 3, 4, 5, 6, 7, 8\}$, and let $A = \{2, 4, 6, 8\}$ and $B = \{1, 2, 3, 4, 5\}$. Write each of the following sets in the form of a list: $A \cup B$, $A \cap B$, $B \backslash A$, $A \backslash B$, A^c, B^c, $(A \cup B)^c$, $(A \cap B)^c$, $(B \backslash A)^c$, $(A \backslash B)^c$, $(A \triangle B)^c$, $(A^c)^c$.

2. Let

$$A = \{x \mid x \in Q \text{ and } 0 < x \leqslant 3\}$$

and let

$$B = \{x \mid x \in R\backslash Q \text{ and } 0 \leqslant x < 3\}.$$

Describe the following sets verbally or by the list or rule methods: $A \cup B$, $A \cap B$, $A\backslash B$, $B\backslash A$, $A \triangle B$.

3. Let the universe of discourse be N, and let A be the set of even integers and B be the set of positive integer powers of two. Describe the following sets verbally or by the list or rule methods:

$$A \cup B, A \cap B, A^c, B^c, A\backslash B, B\backslash A, A \triangle B.$$

4. Identify the sets $Z \cap N$, $Q \cup N$, $R\backslash Q$, $R \triangle N$.

5. Let A and B be two sets. Is there a set X such that $A \cap X = A$ and $B \cap X = B$? Is there a set Y such that $A \cup Y = A$ and $B \cup Y = B$? Are these sets unique?

1.4 SUBSETS

A proper understanding of the use of the symbols to be introduced in this section is closely related to that of the symbols for inequalities of real numbers. You should note that while $3 < 4$ is a true statement, $3 \leqslant 4$ and $3 \leqslant 3$ are also true statements. The truth of statements of the form 'something or something else' will be discussed fully in Chapter 2.

The main object of this section is to build into the language of set theory a notion which corresponds to the observation that of two collections of physical objects, one might be a part of the other. Thus if two people produce lists of their friends, one list might contain every name on the other list. Certainly, if a collection of objects is shared out between several people, than each share is a part of the original collection. In many instances one would expect each share to contain some but not all of the original collection, but this is not always so. For example, when a pack of cards is dealt out for a game of bridge, a hand may contain none or all of the collection of aces in the pack.

The set theoretic notions which reflect these observations are those of subsets and proper subsets. These are introduced in the following way by means of the basic notation for set membership. The symbol $A \subseteq B$ is used, and A is said to be a *subset* of B in the following case and in that case only: if $a \in A$ then $a \in B$. The condition can be expressed in the equivalent form: $a \in A$ implies $a \in B$. One can express it in words by the statement that every element of A is an element of B. If A is a subset of B and if also there is an element of B which does not belong to A, then A is said to be a *proper subset* of B. This is denoted by the symbol $A \subset B$. You will note that if $A \subseteq B$ then either $A \subset B$ or $A = B$. With respect to the statements $A \subseteq B$ and $A \subset B$, both A and B play the role of sets. A Venn diagram to illustrate that A is a subset of B is shown in Fig. 1.5 (a). To indicate that A is a proper subset of B it is necessary to indicate the existence of an element in B which is not in A. This is illustrated in Fig. 1.5 (b).

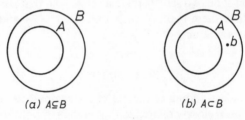

(a) $A \subseteq B$ (b) $A \subset B$

Figure 1.5

Recall that two sets are said to be equal if and only if every element which belongs to one of them belongs to the other. It follows that $A = B$ if and only if $A \subseteq B$ and $B \subseteq A$. Indeed, if two people choose sets independently of each other, then one way to test whether they have chosen the same set or not is to test both $A \subseteq B$ and $B \subseteq A$.

The notions of subset and proper subset can be exemplified by means of the sets $A = \{1, 2\}$, $B = \{1, 2, 3\}$ and $C = \{0, 1, 2\}$. Since $1 \in A$ and $2 \in A$ and A has no other members, and since also $1 \in B$ and $2 \in B$, it follows that $A \subseteq B$. Moreover, since $3 \in B$ and $3 \notin A$, it follows that $A \subset B$. Similarly $A \subseteq C$ and $A \subset C$. On the other hand, since $3 \in B$ and $3 \notin C$, it follows that the statements $B \subseteq C$ and $B \subset C$ are not true. Similarly, since $0 \in C$ and $0 \notin B$, the statements $C \subseteq B$ and $C \subset B$ are not true. Since $A = A$ is a true statement, the statement $A \subseteq A$ is also true. However, the statement $A \subset A$ is false.

If sets A and B have been presented by the rule method, then to test whether or not $A \subseteq B$ one has to appeal to that part of mathematics in which the rules are expressed. Suppose that

$$A = \{x \mid x \in R \text{ and } 0 < x < 7\},$$
$$B = \{x \mid x \in R \text{ and } 0 < x < 5\},$$
$$C = \{x \mid x \in R \text{ and } x^2 < 25\}.$$

Then $A \subseteq B$ is false since $6 \in A$ whilst $6 \notin B$. On the other hand $B \subset C$ is true, since if $0 < x < 5$ then $x^2 < 25$, whilst $-4 \in C$ and $-4 \notin B$.

We have introduced the notion of the empty set $\varnothing$ into the language of sets, and so we must decide for each set A whether or not $\varnothing$ is a subset of A. Intuitively, it would seem that $\varnothing$ should be a subset of every set; but can we show that this follows from our definition of subsets? The set $\varnothing$ is a subset of A if every element of $\varnothing$ is also an element of A. But $\varnothing$ has no elements and so this statement is true! If you wish to disprove that $\varnothing \subseteq A$ you must show that there is an element belonging to $\varnothing$ which does not belong to A, and this you cannot do. You may be unconvinced by these verbal arguments, and if so this provides another instance of the need to formalize the language of proof. We shall show in Chapter 2 that, in terms of this language, for every set A we have $\varnothing \subseteq A$.

In using the language of sets we are often led to consider the collection of subsets of a given set A. This collection is called the *power set of A* and is denoted by $\mathscr{P}(A)$. The name power set is used because of the relationship between the number of elements in a set A and the number of subsets of A (see exercise 7 below). Strangely, we rarely consider the collection of proper subsets of A and no special notation has been agreed on for this. If the rule method of writing sets is to be used to specify $\mathscr{P}(A)$ it must be noted that the elements of $\mathscr{P}(A)$ are subsets of A. In order to help us remember this we shall choose X, rather than x, as the symbol for the variable, and write

$$\mathscr{P}(A) = \{X \mid X \subseteq A\}.$$

Since $A \subseteq A$, it follows that $A \in \mathscr{P}(A)$, and since $\varnothing \subseteq A$, it follows also that $\varnothing \in \mathscr{P}(A)$. Moreover, if $a \in A$ then one may also write $a \in \{a\}$ and $\{a\} \subseteq A$ and $\{a\} \in \mathscr{P}(A)$. The idea of distinguishing between a single object, such as a pen, and a collection of objects

c

consisting solely of that pen, is not one which is ordinarily considered. However, in set theory it is necessary to distinguish between the symbols a and $\{a\}$. To help us to refer to this distinction we introduce the description *singleton set* for a set which has precisely one element. Note that although we refer to $\{a\}$ as a set, it can play the role of an element, as in the statement $\{a\} \in \mathscr{P}(\{a,b,c\})$.

We can now present the power set of the set $A = \{a,b\}$ in the form of a list. The subsets of A are $\varnothing$, $\{a\}$, $\{b\}$ and A, so that

$$\mathscr{P}(A) = \{\varnothing, \{a\}, \{b\}, A\}.$$

The study of power sets can lead far from the everyday world of collections of physical objects. What, for example, should we make of the power set of the empty set? Certainly, since $\varnothing = \varnothing$, it follows that $\varnothing \subseteq \varnothing$ and so that $\varnothing \in \mathscr{P}(\varnothing)$. Similar arguments to those used above to show that $\varnothing$ is a subset of every non-empty set also make it plausible that the empty set has no other subset than itself. This assertion can be shown to be true in terms of the formal language of proof which will be introduced in the next chapter. It follows that $\mathscr{P}(\varnothing) = \{\varnothing\}$. It is interesting to note that $\{\varnothing\}$ is a singleton set whose only 'element' is the empty 'set', and so is not equal to the empty set $\varnothing$ itself, which leads to the result that the power set of the empty set is not empty. Consequently, $\varnothing \subseteq \{\varnothing\}$, the empty set being a subset of every set, and also $\varnothing \in \{\varnothing\}$, because of the rules for the use of the symbol $\in$ with sets presented by the list method. The only subsets of the set $\{\varnothing\}$ are therefore $\varnothing$ and $\{\varnothing\}$, and it follows that

$$\mathscr{P}(\mathscr{P}(\varnothing)) = \mathscr{P}(\{\varnothing\}) = \{\varnothing, \{\varnothing\}\}.$$

The common names 'element', 'set' and 'empty set' do not help in this situation. One must use the notations of set theory in accordance with its rules.

Exercises 1.4

1. Let $A = \{1,2,3,4\}$, $B = \{2,3,4,5,6\}$ and $C = \{2,4\}$. State which of the following are true: $A \subseteq B$, $B \subseteq A$, $C \subseteq B$, $B \subseteq C$, $C \subseteq A$, $C \subset A$, $C \subset B$, $B \subset C$, $C \in A$, $\varnothing \subset B$.

2. State which of the following relationships are true. $Q \subseteq R$, $Q \subset R$, $R \subset Z$, $Z \subseteq N$, $N \subseteq Z$, $Z \subseteq R$.

3. Determine whether the following statements are true or false.
   (a) If $A \subseteq B$ and $B \subseteq C$ then $A \subseteq C$.
   (b) If $A \subseteq B$ and $B \nsubseteq C$ then $A \nsubseteq C$.

4. Let $A = \{a, b, c, d\}$. Write the set $\mathcal{P}(A)$ in the form of a list.

5. Let $A = \{a, b, \{a\}\}$. Write the set $\mathcal{P}(A)$ in the form of a list.

6. Let $A = \{\{a\}, b\}$, $B = \{\{a, b\}, a\}$ and $C = \{a\}$. State which of the following are true: $A \subset B$, $C \subset A$, $C \in A$, $C \subset B$, $C \in \mathcal{P}(A)$, $C \in \mathcal{P}(B)$, $\varnothing \in \mathcal{P}(A)$, $\varnothing \subseteq \mathcal{P}(A)$, $\{\varnothing\} \in \mathcal{P}(A)$, $\{\varnothing\} \subseteq \mathcal{P}(A)$.

7. If the set A has n elements, where n is a natural number, how many elements has $\mathcal{P}(A)$? Decide on a method of proof of the result.

1.5 RELATIONSHIPS INVOLVING OPERATIONS

There are a number of important relationships between sets constructed by means of the operations described in §1.3. In this section we shall extend the use of Venn diagrams to illustrate and investigate these relationships. We would emphasize at this point that in our philosophy of what constitutes a mathematical proof, the validity of such relationships can be made merely plausible by the use of a Venn diagram; they cannot be *proved* in this way. One reason which leads us to adopt this standpoint is that the relationships may involve sets whose elements are not points in the plane, and so in

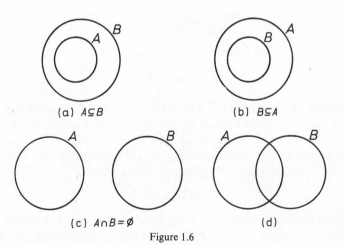

(a) $A \subseteq B$ (b) $B \subseteq A$

(c) $A \cap B = \varnothing$ (d)

Figure 1.6

using plane regions to illustrate them we may be making unsuspected changes in the relationships. Even after having seen that Venn diagrams do illustrate many true relationships between other kinds of sets, we do not allow them as methods of proof. To do so would be to permit of a process analogous to induction in the physical sciences, and our philosophy of what constitutes a mathematical proof does not admit this as a valid process. Further discussion of the limitations of diagrams, and of mathematical proof, will comprise the next chapter.

We shall begin the discussion of relationships involving operations with a consideration of the set $A\backslash(A\backslash B)$. There are essentially four different configurations for the regions representing the sets A and B in a Venn diagram. These are shown in Fig. 1.6.

The configuration of Fig. 1.6 (d) is the most general one in the sense that under suitable restrictions it acts like each of the other configurations. For example, if $A\backslash B = \varnothing$ then Fig. 1.6 (d) acts like Fig. 1.6 (a). To study $A\backslash(A\backslash B)$ we use the most general configuration for the sets A and B, with the diagram suitably shaded. In Fig. 1.7 the region representing A is shaded with horizontal lines, whilst that representing $A\backslash B$ is shaded with vertical lines. The collection

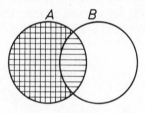

Figure 1.7

of points belonging to A and not to $A\backslash B$ would appear to be represented by the region containing only horizontal shading, and from Fig. 1.2 this corresponds to $A\cap B$. Since every other possible diagram will be essentially the same as Fig. 1.6 (d) or a special case of it, it appears plausible that for every pair of sets A and B,

$$A\backslash(A\backslash B) = A\cap B.$$

As a further example we may illustrate the set $A\backslash B$ within a universe of discourse U, again using the most general configuration.

Consideration of the diagrams illustrating complement and intersection together with Fig. 1.8 then makes the statement

$$A\backslash B = A \cap B^c,$$

appear plausible for every pair of sets A and B.

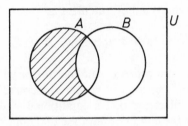

Figure 1.8

We have discussed the plausible inference of relationships from diagrams. One can work also in the other direction and use diagrams to make it plausible that some relationships are not generally valid. In exercise 1 below you are asked to draw Venn diagrams to investigate given relationships to decide whether or not they are generally valid. Venn diagrams can also be used to remind one of particular valid relations. For example in exercise 2 below one of the symbols has been replaced by a question mark, and you can use Venn diagrams to decide what the symbol should be.

Venn diagrams can be used in situations in which more than two sets are involved. The most general configuration for three sets is

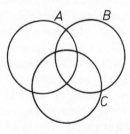

Figure 1.9

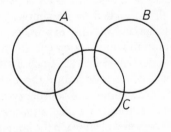

Figure 1.10

illustrated in Fig. 1.9. If necessary, the diagram can be enclosed in a region which represents the universe of discourse. In special cases one or more of the seven regions in the diagram may be empty. For example if $A \cap B = \varnothing$ then the diagram can take the special form of Fig. 1.10.

So far we have only considered the use of Venn diagrams to show that certain statements are plausible for all sets A and B. An example of another kind of statement is

'if $A \subseteq B$ then $A \cap B = A$'.

Such statements are called *conditional statements*, or *statements of implication*. They consist of two parts, a condition followed by a conclusion. To illustrate that a conditional statement is plausible one must build the condition into the diagram. Now the condition $A \subseteq B$ is illustrated in Fig. 1.6 (a), and for that figure $A \cap B$ is just A. Consideration of the same figure makes plausible the statement

'if $A \subseteq B$ then $A \cup B = B$'.

Similarly consideration of Fig. 1.10 makes plausible the statement

'if $A \cap B = \varnothing$ then $(A \cap C) \cap (B \cap C) = \varnothing$'.

The notations of set theory, and the rules for their use, including the relationships discussed in this section, have been introduced to provide a mathematical model of everyday ideas of elements and sets. The model is designed to be a good fit for a wide range of situations, such as those which involve no more than the sorting of a set of pencils on a desk into piles of pencils of the same colour. The model is also designed to be simple, unambiguous and powerful. Now consider an everyday expression involving sets and elements which is an ambiguous statement about a complex situation such as has been described in §1.2. Such a situation cannot be investigated by means of the model until the ambiguities have been resolved. Even then little enough of set theory is likely to be needed to study the most complex everyday situation. The great power of the model only becomes clear when the language of sets is used in other branches of mathematics.

Exercises 1.5

1. Use Venn diagrams to determine whether or not it is plausible that the following statements are valid for all sets A, B and C.

(a) $A \cap (A \backslash B) = A$.

(b) $A \triangle (B \backslash A) = A \cup B$.

(c) $A \backslash (B \cup A) = B$.

(d) $A \cap (B \backslash C) = (A \cap B) \backslash (A \cap C)$.

(e) $A \triangle (B \backslash C) = (A \triangle B) \backslash (A \triangle C)$.

(f) $A \cup (B \backslash C) = (A \cup B) \backslash (A \cup C)$.

(g) If $A \cap B = \varnothing$ then $A \triangle B = A \cup B$.

(h) If $A^c = B$ then $A \backslash B = B$.

(i) If $C = A \triangle B$ then $A \cup (B \cap C) = A \cup B$.

2. Use Venn diagrams to determine whether there are symbols which can be used to replace the question marks in the statements below so as to make it plausible that they are valid for all sets A, B and C.

(a) $(A \cup B)^c = A^c \, ? \, B^c$.

(b) $A \backslash (B \, ? \, A) = A$.

(c) If $A = B \cup C$ then $(A \backslash C) \, ? \, B = B$.

(d) If $A = B \backslash C$ then $B \backslash A = B \, ? \, C$.

2

MATHEMATICAL PROOF

2.1 THE LIMITATIONS OF DIAGRAMS

In considering the examples of §1.5, we emphasized that the diagrams made certain statements plausible, but that consideration of the diagrams could not be regarded as mathematical proof of the general validity of the statements.

In that section the most complicated type of statement which was considered was a conditional statement of the form 'if $A \subseteq B$ then $A \cap B = A$'. In this case it is easy to believe that if two people independently draw diagrams to illustrate the statement, then the two diagrams will be substantially the same, and moreover will be substantially the same as any other diagram which can be produced. However, the conditional statements which are investigated in mathematics can be much more involved than this. The condition might have several parts, rather than just one such as $A \subseteq B$, and the conclusion might also be much more involved than $A \cap B = A$. In these circumstances two diagrams illustrating the situation might be substantially different, and it would not be at all clear that for every possible diagram for which the conditions of the statement held, the conclusion would also hold.

As an example of a complicated statement, consider the statement $B \cap D \subseteq (A \cap B) \cup (B \cap C)$. If the configuration of Fig. 2.1(a) is

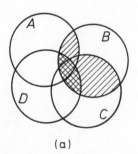

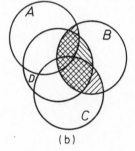

(a)   (b)

Figure 2.1

drawn then one concludes that this statement might be true for every collection of four sets A, B, C and D. However, if the configuration of Fig. 2.1(b) is drawn then it is seen that the statement is not universally true. Neither of the configurations is the most general one for four sets. Indeed the most general configuration for four sets cannot be constructed with four circles. However, you will be able to construct it by using four rectangles. It is of limited use because of the large number of regions which it contains.

It must be noted that the use of Venn diagrams is limited to situations which involve the operations of sets, and inclusion relationships. They illustrate these aspects but not others. For example, Fig. 2.2 illustrates the inclusion $Q \subseteq R$ of the rationals within the reals, but the geometry of the diagram is wholly misleading concerning these sets on the number line. It would not help one to decide whether the following statement about numbers is plausible: between every pair of real numbers there is a rational number. A diagram of points on a number line might assist towards a proof of this statement, but such diagrams are limited in that they cannot represent the relationship between the set of rational numbers and the set of irrational numbers.

Figure 2.2

Finally, the use of diagrams is limited to statements for which one can envisage some kind of diagram, and for many mathematical statements one can not. For example, it is difficult to imagine any diagram which would assist with a proof of the statement that n^2 is an even integer if and only if n is an even integer.

2.2 PROOF

We have seen in the examples above that something more than diagrams is required to provide proofs of mathematical statements.

We also saw that the kind of verbal argument about the empty set used in § 1.3 and § 1.4 gives rise to questions about what constitutes a mathematical proof. Thus there seems to be a need for some discussion of what is involved in proof.

We would emphasize at this point that 'proof' depends very much on the context in which it is used. The ideas of procedures of proof are different for a physicist, an archaeologist, a lawyer and a mathematician. We would also emphasize that proof is a function of time, and to illustrate this in mathematics would point to the development of the idea of proof in the calculus from the time of Newton to the present day. This naturally implies that proof may be different in different social context; and in future times.

Our object in this chapter is to lead the reader to a greater understanding and ability in dealing with mathematical proofs as they are usually presented by present-day mathematicians. As a means to that end we shall build a system which models the processes of mathematical proof as we see them. Some proofs will then be presented in two forms, one of which displays explicitly the relationship with our symbolic description of proof, in order to throw light on the usual form in which proofs are presented. We regard the symbolic description of proof as transitional in the sense that once its object has been achieved it will no longer be employed explicitly.

We shall use mathematical terminology in our description of mathematical proof. The situation is similar to that which arises when the English language is used to describe the structure of the English language. However, some people have preferred to try to base the whole of mathematics on the study of logic in order to avoid any possible circularities of argument.

2.3 SYMBOLIC LANGUAGE

We shall begin our discussion by referring back to two examples from Chapter 1, and shall analyse in detail the content of these examples to see how the language used can be symbolized. These examples, which were shown to be plausible by diagrams in § 1.5, involved the statements

  (i) given any pair of sets A and B, $A \backslash B = A \cap B^c$,
  (ii) if $A \subseteq B$ then $A \cup B = B$.

The first statement is taken to mean that the set which the symbol $A \backslash B$ represents is equal to the set denoted by the symbol $A \cap B^c$. We must therefore look behind the symbols to the sets they represent. Recall that B^c is the complement of B with respect to some fixed universe of discourse U, and that $x \in B^c$ is an abbreviation for '$x \in U$ and $x \notin B$'. The notation $A \cap B^c$ stands for the set $\{x \mid x \in A$ and $x \in B^c\}$ which is therefore the set $\{x \mid x \in A$ and $x \notin B\}$, it being understood that $x \in U$. So it would indeed appear that $A \cap B^c$ and $A \backslash B$ are two different symbols for the same set.

The symbols which occur in statement (ii) are also abbreviations of longer statements. It was seen in §1.4 that $A \subseteq B$ is an abbreviation for '$x \in A$ implies $x \in B$', and in §1.1 that $A \cup B = B$ is an abbreviation for '$x \in A \cup B$ if and only if $x \in B$', while $x \in A \cup B$ itself stands for '$x \in A$ or $x \in B$'. Breaking up the statement in this way, it can be rephrased in terms of $x \in A$, $x \in B$ and connecting words as follows:

(iii) if $x \in A$ implies $x \in B$ then $x \in A$ or $x \in B$ if and only if $x \in B$.

Similarly the statement (i) can now be put in the form '$x \in A$ and $x \notin B$ and $x \in U$ if and only if $x \in A$ and $x \in U$ and $x \notin B$'. This can be expressed as follows in terms of $x \in A$, $x \in B$ and connecting words if 'not $x \in B$' is used for '$x \notin B$':

(iv) $x \in A$ and not $x \in B$ and $x \in U$ if and only if $x \in A$ and $x \in U$ and not $x \in B$.

Proofs of the above statements will now involve a consideration of the logical structure of statements like (iii) and (iv), which have been constructed out of basic statements like $x \in A$, $x \in B$ with various connecting words such as, 'not', 'and', 'or', 'if-then', 'implies', 'if and only if'.

We shall consider these connecting words applied to general statements, for which letters p, q, r, etc. will be used. First the word 'not'. The negation of the statement 'set theory is easy' is 'set theory is not easy'. You could say 'not—set theory is easy' but probably would not. We have agreed that the notation for the negation of '$x \in A$' is to be '$x \notin A$', but we could write 'not $x \in A$'. As a notation for the negation of a general statement p, we choose $\sim p$. One cannot tell in general whether p or $\sim p$ is a true statement. Some of you may think that the statement 'set theory is easy' is true and some of you will think that it is false. But certainly those who think it true will think that the negation of it is false, and those who think it false will think that the negation of it is true. This will be the case for each pair of statements p and $\sim p$ which are used in the mathematical system of proof described in this chapter. One of the two statements will be true and the other false, though which of them is true will depend on the context. In contrast, in Chapter 5 we shall ascribe probabilities to some statements rather than declare them to be true or false.

Now if p and q are statements, we shall allow 'p and q' as a statement and denote it by $p \wedge q$. It will be called the *conjunction* of p and q. Just as the truths of the statements p, $\sim p$ are related, so the truths of the statements p, q, $p \wedge q$ are related. If p and q are both known to be true then $p \wedge q$ is true, while if either of them is known to be false, then $p \wedge q$ is false.

In everyday language no ambiguities occur with the use of the word 'and', and provided care is taken over double negatives no ambiguities need arise with the use of the word 'not'. Thus there is no problem over the use of these words in mathematics. However, in everyday language the word 'or' is used in the non-exclusive sense of 'and/or', in the exclusive sense of 'but not both', and ambiguously. Since mathematical language needs to be unambiguous, we distinguish between the two senses by using 'p or q' in the sense of 'and/or', and by using 'either p or q' in the sense of 'but not both'. Thus for any pair of statements p and q, we shall allow 'p or q' as a statement and denote it by $p \vee q$. It will be called the *disjunction* of p and q. If p and q are both known to be false then $p \vee q$ is false; otherwise $p \vee q$ is true. There is no universally adopted symbol for the statement 'either p or q'. It is regarded formally as an abbreviation for 'p or q, and not p and q', which can be expressed in terms of the symbols $\sim$, $\wedge$ and $\vee$ (see exercise 4 of §2.4).

In order to summarize the uses of the connectives given above and

TABLE 2.1 Truth tables of (a) negation (b) conjunction (c) disjunction

(a) p	$\sim p$	(b) p	q	$p \wedge q$	(c) p	q	$p \vee q$
T	F	T	T	T	T	T	T
F	F	T	F	F	T	F	T
		F	T	F	F	T	T
		F	F	F	F	F	F

obtain a convenient form in which to present them we change our language slightly. A statement which is known to be true will be said to have truth value T, and a statement which is known to be false will be said to have truth value F.

The related truth values of p, q, $\sim p$, $p \wedge q$ and $p \vee q$ are shown in rows under the appropriate headings in Table 2.1 (a), (b) and (c), which are called the truth tables of negation, conjunction and disjunction, respectively.

So far we have dealt with the connecting words 'not', 'and', 'or'. Now we must turn to words which denote equivalence and implication. There is no ambiguity about the way in which the phrase 'if and only if' is used in everyday language. The statement 'p if and only if q' is generally agreed to be true when p and q have the same truth value, and false when they have different truth values. We therefore adopt the following truth table, denoting the connective 'if and only if' by the symbol $\Leftrightarrow$.

TABLE 2.2   Truth table for 'if and only if'

p	q	$p \Leftrightarrow q$
T	T	T
T	F	F
F	T	F
F	F	T

The two phrases denoting implication, 'p implies q' and 'if p then q', mean the same thing and they are expressed in symbols by the statement '$p \Rightarrow q$'. It is not so easy to decide what the truth table for $p \Rightarrow q$ should be, as it was for negation, conjunction, disjunction and equivalence. According to everyday usage, if p is true and q is true then we accept that $p \Rightarrow q$ is true, while if p is true and q is false, we accept that $p \Rightarrow q$ is false. Thus we have the following truth table so far.

TABLE 2.3   Incomplete truth table for implication

p	q	$p \Rightarrow q$
T	T	T
T	F	F
F	T	?
F	F	?

It is not so clear what truth values should be ascribed to $p \Rightarrow q$ when p is false. However, if the system we are constructing is to be of any use to us in ascribing truth values to complicated statements, then we certainly cannot leave the last two entries blank. Consideration of a variety of statements used in everyday language does not really clarify things. The statement 'if I catch the early train then I shall be in time for lunch' carries the strong implication that if I miss the early train then I shall be late for lunch. However, the statement 'if John comes to your party, then you will receive a present' would probably merely leave you in some doubt as to whether or not you would receive a present if he did not come. Although this type of statement does not give direct guidance as to how the truth table for $p \Rightarrow q$ should be completed, it does indicate that the statement 'p implies q' is used in a different sense from the statement 'q' and from the statements 'p and q' and 'p if and only if q'. The truth table for $p \Rightarrow q$ should therefore be different from the truth tables for q and for $p \wedge q$ and for $p \Rightarrow q$. In all four cases the first two entries are TF. Of the four possible ways of completing the last two entries three are used by q, $p \wedge q$ and $p \Leftrightarrow q$ leaving for $p \Rightarrow q$ the entries TT (see Table 2.4). The truth table for implication completed in this way will apply to all mathematical statements of the form $p \Rightarrow q$. As we have already seen, some everyday usages are contrary to this and some are merely ambiguous.

We shall now summarize the connectives and their truth tables. Negation is given by Table 2.1(a), and the other connectives are collected together in Table 2.4. These two tables can be regarded as a model of the logic of everyday language. Most of the entries are derived from unambiguous everyday usages of language, and the

TABLE 2.4   Summary of truth tables

p	q	$p \wedge q$	$p \vee q$	$p \Leftrightarrow q$	$p \Rightarrow q$
T	T	T	T	T	T
T	F	F	T	F	F
F	T	F	T	F	T
F	F	F	F	T	T

remaining ones are chosen in such a way as to produce a complete and unambiguous framework for the logic of mathematical language.

Words such as 'if', 'necessary', 'sufficient' are also used imprecisely in everyday speech in connection with implication. In mathematical language we require more precise interpretations, and so we summarize mathematical usage below in connection with our symbolization.

TABLE 2.5   Phrases denoting implication

$p \Rightarrow q$	$q \Rightarrow p$
p implies q	q implies p
if p then q	if q then p
q if p	p if q
p only if q	q only if p
p is sufficient for q	q is sufficient for p
q is necessary for p	p is necessary for q
q follows from p	p follows from q

Exercises 2.3

1. Analyse the logical structure of the following statements by expressing them in symbolic language. Compare the ways in which you have formalized the statements with the ways in which other people in your class have done so.

(a) If you attach a rope to the car, release the brake and pull hard, then the car will begin to move or the rope will break.

(b) If when driving you feel drowsy then you should immediately stop and rest, or open the window and admit some fresh air.

(c) The equation $ax^2 + bx + c = 0$ will not have any real roots unless $b^2 - 4ac \geqslant 0$.

(d) You will not find your way if you do not buy a map and plan your route carefully.

(e) Foreign travel is only worthwhile if you are prepared to try local food and to investigate local customs.

2. Let p stand for the statement 'I work hard'.
   Let q stand for the statement 'I enjoy mathematics'.
   Let r stand for the statement 'I am happy'.

Translate the following symbolic statements into English.

(a) $p \wedge r$.   (c) $\sim r \Leftrightarrow (p \wedge \sim q)$.   (e) $q \Rightarrow (p \vee \sim r)$.

(b) $p \Rightarrow (q \vee r)$.   (d) $p \wedge (q \vee r)$.

2.4 DERIVED TRUTH TABLES AND TAUTOLOGIES

In the previous section we considered the truth tables for the simplest compound statements, $\sim p$, $p \wedge q$, $p \vee q$, $p \Leftrightarrow q$, $p \Rightarrow q$. All other compound statements are constructed from these, and the truth tables are built up from the truth tables in the last section.

Let us begin by looking again at the statement $p \Leftrightarrow q$. In common usage, 'p if and only if q' means the same as 'p implies q and q implies p'. We shall investigate whether our symbolism is consistent with this by constructing the truth table for the compound statement $(p \Rightarrow q) \wedge (q \Rightarrow p)$.

TABLE 2.6

p	q	$p \Rightarrow q$	$q \Rightarrow p$	$(p \Rightarrow q) \wedge (q \Rightarrow p)$
T	T	T	T	T
T	F	F	T	F
F	T	T	F	F
F	F	T	T	T

The entries in columns 3 and 4 of Table 2.6 are found by reference to Table 2.4, whilst column 5 is obtained from columns 3 and 4 using Table 2.1(b). We now see that the statement $(p \Rightarrow q) \wedge (q \Rightarrow p)$ has indeed the same truth table as $p \Leftrightarrow q$, and this adds further to our conviction that we have chosen a sensible truth table for implication. It also shows that if we wish we can dispense with $\Leftrightarrow$ by expressing it in terms of $\Rightarrow$ and $\wedge$. The interested reader can investigate whether any further reduction can be made in the number of connectives needed.

Truth tables for statements which are built up from two basic statements p, q by use of the connectives $\sim$, $\wedge$, $\vee$, $\Leftrightarrow$, $\Rightarrow$, will not

TABLE 2.7 Truth table for $(q \vee r) \Rightarrow (p \wedge \sim r)$

p	q	r	$\sim r$	$q \vee r$	$p \wedge \sim r$	$(q \vee r) \Rightarrow (p \wedge \sim r)$
T	T	T	F	T	F	F
T	T	F	T	T	T	T
T	F	T	F	T	F	F
T	F	F	T	F	T	T
F	T	T	F	T	F	F
F	T	F	T	T	F	F
F	F	T	F	T	F	F
F	F	F	T	F	F	T

be too difficult to draw up. They might need a large number of columns, but they will need only four rows to deal with the four different possible pairs of truth values for p and q. However, for n basic statements there are 2^n possible sets of initial truth values, and so a truth table with 2^n rows is required. Even when $n = 4$, this will be cumbersome to handle. An example of a truth table for a statement which involves three basic statements p, q, r is given above.

This table shows that if the basic statements p, q, r have the truth value T then a compound statement need not have the truth value T. Similar examples for compound statements of two basic statements can be given. For example, when p and q each have the truth value T the compound statement $\sim p \wedge q$ has the truth value F. Compound statements which do have the truth value T when all the basic statements from which they are composed have the truth value T are particularly important. They are said to be *true relative to the basic statements*. Thus the statement $17^4 > 5^7$, which by a process of calculation you can easily convince yourself to be true, will be accepted as true relative to certain basic statements about arithmetic which embody the techniques which you will have used in your calculation. For another example we take arithmetic modulo 2. The numbers for the arithmetic are just 0 and 1, and addition of two numbers is as for the ordinary integers except that $1 + 1 = 0$. Now the statement $1 + 1 + 1 = 1$ is true relative to the basic statements of arithmetic modulo 2, but of course it is false relative to the basic statements of ordinary arithmetic.

Some compound statements have the truth value T no matter what the initial truth values of the basic statements. Such compound statements are called *tautologies*. The simplest tautological statement is 'p or not p', i.e. $p \vee \sim p$; for if p is true then $p \vee \sim p$ is true, whilst if p is false then $\sim p$ is true so that $p \vee \sim p$ is true.

Two tautologies which are used constantly in arguments are $p \wedge q \Rightarrow p$ and $p \Rightarrow p \vee q$. The truth tables are given below.

TABLE 2.8   Tautological implication $p \wedge q \Rightarrow p$

p	q	$p \wedge q$	$p \wedge q \Rightarrow p$
T	T	T	T
T	F	F	T
F	T	F	T
F	F	F	T

D

TABLE 2.9   Tautological implication $p \Rightarrow p \vee q$

p	q	$p \vee q$	$p \Rightarrow p \vee q$
T	T	T	T
T	F	T	T
F	T	T	T
F	F	F	T

The first of these tautologies embodies the logic behind a claim such as 'x is a real number and x is positive implies x is a real number', a remark which you are not likely to dispute. The second, however, embodies the logic behind such statements as $3 \leqslant 4$. Suppose p stands for the statement '$x < 4$' and q stands for the statement '$x = 4$'. Then $p \vee q$ stands for the statement '$x < 4$ or $x = 4$', which is commonly abbreviated to '$x \leqslant 4$'. In this case the tautology $p \Rightarrow p \vee q$ becomes '$x < 4$ implies $x \leqslant 4$'. In particular it follows that $3 \leqslant 4$ is a true statement.

These two tautologies are of the form $P \Rightarrow Q$ where P and Q are expressed in terms of basic statements p and q. If $P \Rightarrow Q$ is true for all possible truth values of p and q, then we say that P *tautologically implies* Q. If P tautologically implies Q and Q tautologically implies P, so that the statement $P \Leftrightarrow Q$ is a tautology, we say that P *is tautologically equivalent to* Q. This will only be the case when, whatever the truth values for the basic statements p, q, the statements P, Q have the same truth value. So in any logical argument the statement P can be replaced by Q without the truth values involved being affected. As a first simple example of this, we give the truth table which will confirm the intuitive notion that statements such as 'he is a doctor and he is married' and 'he is married and he is a doctor' are logically equivalent. The statements are of the form $p \wedge q$, $q \wedge p$, and the truth table below shows that $(p \wedge q) \Leftrightarrow (q \wedge p)$ is a tautology.

TABLE 2.10   Tautological equivalence $(p \wedge q) \Leftrightarrow (q \wedge p)$

p	q	$p \wedge q$	$q \wedge p$	$(p \wedge q) \Leftrightarrow (q \wedge p)$
T	T	T	T	T
T	F	F	F	T
F	T	F	F	T
F	F	F	F	T

We noticed at the beginning of this section that the truth table for $\Leftrightarrow$ can be built up from those for $\Rightarrow$ and $\wedge$. We can also build up the truth table for implication in terms of other connectives. Consider the statement $(p \Rightarrow q) \Leftrightarrow (\sim p \vee q)$, whose truth table is obtained below.

TABLE 2.11

p	q	$\sim p$	$\sim p \vee q$	$p \Rightarrow q$	$(p \Rightarrow q) \Leftrightarrow (\sim p \vee q)$
T	T	F	T	T	T
T	F	F	F	F	T
F	T	T	T	T	T
F	F	T	T	T	T

The given statement is therefore a tautology, so that $p \Rightarrow q$ is tautologically equivalent to $\sim p \vee q$. This situation is explored further in exercise 5 below.

In many pieces of mathematical work one wishes to show that a statement p does not imply a statement q, and the following discussion deals with the corresponding situation in the logical model. The truth table for the statement $\sim(p \Rightarrow q)$ is given in Table 2.12, from which it will be seen that the statement $\sim(p \Rightarrow q)$ has truth value T if and only if p has truth value T and q has truth value F. This suggests that the statement $\sim(p \Rightarrow q)$ might be equivalent to the statement $p \wedge \sim q$, and this is confirmed by means of Table 2.13. Since the statement $\sim(p \Rightarrow q)$ has the same truth table as the statement $p \wedge \sim q$ they are tautologically equivalent. This can be related to

	TABLE 2.12					TABLE 2.13		
p	q	$p \Rightarrow q$	$\sim(p \Rightarrow q)$		p	q	$\sim q$	$p \wedge \sim q$
T	T	T	F		T	T	F	F
T	F	F	T		T	F	T	T
F	T	T	F		F	T	F	F
F	F	T	F		F	F	T	F

the result of Table 2.11. From that table the statements $p \Rightarrow q$ and $\sim p \vee q$ are equivalent, and so their negations will be equivalent, i.e. $\sim(p \Rightarrow q)$ is equivalent to $\sim(\sim p \vee q)$. The use of de Morgan's law (tautology (xv) below) confirms that $\sim(\sim p \vee q)$ and $p \wedge \sim q$ are in fact equivalent statements.

At this point we recall that a compound statement Q is said to be true relative to basic statements $p_1, p_2, \ldots, p_n$, if it has truth value T whenever all the basic statements have truth value T. When each one of the statements $p_1, p_2, \ldots, p_n$ has truth value T, the compound statement $p_1 \wedge p_2 \wedge \ldots \wedge p_n$ will have truth value T. If Q also has truth value T, then the statement $(p_1 \wedge p_2 \wedge \ldots \wedge p_n) \Rightarrow Q$ will have truth value T. If one or more of the basic statements has truth value F, then $p_1 \wedge p_2 \wedge \ldots \wedge p_n$ will have truth value F, and so again the statement $(p_1 \wedge p_2 \wedge \ldots \wedge p_n) \Rightarrow Q$ will have truth value T, and will thus be a tautology. Conversely, if $(p_1 \wedge p_2 \wedge \ldots \wedge p_n) \Rightarrow Q$ is a tautology, then Q will have truth value T when each of the basic statements $p_1, p_2, \ldots, p_n$ has truth value T. We can summarize this discussion by saying that the compound statement Q is true relative to basic statements $p_1, p_2, \ldots, p_n$ if and only if the statement $(p_1 \wedge p_2 \wedge \ldots \wedge p_n) \Rightarrow Q$ is a tautology.

Tautologies commonly used in logical arguments are listed below. Some of them are tautological implications and some are tautological equivalences. The names by which they are commonly known are also given.

All the tautologies in the list are used explicitly in formal presentations of proofs, and implicitly in the ordinary discursive presentations of proofs which are used in most mathematics texts. However, only a few of them are referred to explicitly in discursive presentations of proofs. Of these the most common are tautologies (xvii) and (xviii), and less frequently (x), (xv) and (xvi). Tautologies (vi) to (ix), (xiii) and (xiv) tend to be used consciously by mathematicians even though they are not explicitly referred to in written work. The rest of the tautologies tend to be used unconsciously by mathematicians.

Tautologies

   (i) $(p \wedge q) \Leftrightarrow (q \wedge p)$

      $(p \vee q) \Leftrightarrow (q \vee p)$ (commutative laws)

  (ii) $[p \wedge (q \wedge r)] \Leftrightarrow [(p \wedge q) \wedge r]$

      $[p \vee (q \vee r)] \Leftrightarrow [(p \vee q) \vee r]$ (associative laws)

(iii) $[p \wedge (q \vee r)] \Leftrightarrow [(p \wedge q) \vee (p \wedge r)]$
      $[p \vee (q \wedge r)] \Leftrightarrow [(p \vee q) \wedge (p \vee r)]$ (distributive laws)

(iv) $p \wedge p \Leftrightarrow p$
     $p \vee p \Leftrightarrow p$ (idempotent laws)

(v) $p \Leftrightarrow \sim(\sim p)$ (law of double negation)
(vi) $p \Rightarrow p \vee q$
     $p \Rightarrow (q \Rightarrow p)$ (laws of addition)

(vii) $p \wedge q \Rightarrow p$ (law of simplification)
(viii) $[p \wedge (p \Rightarrow q)] \Rightarrow q$ (law of detachment)
(ix) $[(p \Rightarrow q) \wedge (q \Rightarrow r)] \Rightarrow (p \Rightarrow r)$
     $[(p \Leftrightarrow q) \wedge (q \Leftrightarrow r)] \Rightarrow (p \Leftrightarrow r)$ (transitive laws)

(x) $[(p \Rightarrow q) \wedge (r \Rightarrow q)] \Leftrightarrow [p \vee r \Rightarrow q]$ (proof of cases)
   $[(p \Rightarrow q) \wedge (p \Rightarrow r)] \Leftrightarrow [p \Rightarrow q \wedge r]$ (proof by parts)

(xi) $(p \Rightarrow q) \Rightarrow [(r \wedge p) \Rightarrow (r \wedge q)]$ (addition to implication)
    $(p \Rightarrow q) \Rightarrow [p \Rightarrow (p \wedge q)]$

(xii) $\frown(p \wedge \frown p)$
     $p \vee \sim p$ (laws of contradiction)

(xiii) $(p \Rightarrow q) \Leftrightarrow (\sim p \vee q)$

(xiv) $(p \wedge \sim p) \Rightarrow q$
     $[(p \vee q) \wedge \sim p] \Rightarrow q$

(xv) $\sim(p \wedge q) \Leftrightarrow (\sim p \vee \sim q)$
    $\sim(p \vee q) \Leftrightarrow (\sim p \wedge \sim q)$ (de Morgan's laws)

(xvi) $(p \Rightarrow q) \Leftrightarrow (\sim q \Rightarrow \sim p)$ (law of contrapositive)
(xvii) $[(p \Rightarrow q) \wedge (p \Rightarrow \sim q)] \Rightarrow \sim p$ (law of absurdity)
(xviii) $[(p \wedge \sim q) \Rightarrow (r \wedge \sim r)] \Leftrightarrow (p \Rightarrow q)$ (*reductio ad absurdum*)

Exercises 2.4

1. Write out the truth tables for the following compound statements and state which of them are tautologies.

(a) $(q \wedge p) \Rightarrow (\sim p \vee q)$.   (e) $p \Rightarrow q \vee r$.

(b) $\sim(p \wedge q) \Rightarrow p$.   (f) $[(p \wedge q) \vee r] \Rightarrow (p \vee r)$.

(c) $(p \Rightarrow q) \Leftrightarrow (q \Rightarrow p)$.   (g) $(\sim p \vee q) \Leftrightarrow (r \wedge p)$.

(d) $[p \wedge (p \Leftrightarrow q)] \Rightarrow q$.   (h) $[(p \Leftrightarrow q) \wedge (q \Rightarrow r)] \Rightarrow (p \Rightarrow r)$.

2. Confirm that the compound statements in the list above are indeed tautologies.

3. Find some tautologies, involving two or three basic statements, which do not occur in the list of tautologies given above.

4. Construct what you consider to be a suitable truth table for the statement 'either p or q'. Show that this statement can be expressed by means of the connectives $\sim$, $\wedge$, $\vee$.

5. Use the results of Tables 2.6 and 2.11 to express $p \Leftrightarrow q$ in terms only of the connectives $\sim$, $\wedge$, $\vee$.

6. Investigate the connective $|$ defined by $p \,|\, q = \sim(p \wedge q)$ to see how many of the other logical connectives can be expressed in terms of it.

2.5 STATEMENTS INVOLVING VARIABLES

Consider the two statements '$17^4 - 5^7 > 0$' and '$x > 0$'. The first is true relative to the basic statements of ordinary arithmetic. On the other hand, there is no single truth value for the statement $x > 0$; the truth value depends on the variable x. In each context in which the statement is used, some universe of discourse will have been specified (or at least understood). If the universe of discourse is taken to be $\{-1, 0, 1\}$ in this case, then after substituting special values for the variable x, the three statements $-1 > 0, 0 > 0, 1 > 0$ are obtained. The first two of these are false, and the third is true, in ordinary arithmetic.

A similar situation occurs for every statement $P(x)$ built up from basic statements and involving a variable x. There is no single truth table for $P(x)$. Rather, there is a truth table for each value of x taken from the universe of discourse U. In this case $P(x)$ may be called an *open statement* or formula, in contrast with a compound statement with a single truth table which may be called a *closed statement*. If a particular value $x_0 \in U$ is substituted for the variable x in an open statement $P(x)$, then a closed statement $P(x_0)$ is obtained.

If $P(x)$ and $Q(x)$ are open statements, then $\sim P(x_0)$, $P(x_0) \wedge Q(x_0)$ and $P(x_0) \vee Q(x_0)$ are defined for each element x_0 in the universe of discourse U for x. The notations $\sim P(x)$, $P(x) \wedge Q(x)$ and $P(x) \vee Q(x)$

will be used for the open statements from which the corresponding closed statements are obtained by the substitution of specific elements for the variable x. The notations $P(x) \Leftrightarrow Q(x)$ and $P(x) \Rightarrow Q(x)$ are used in a similar way. In substituting a specific element x_0 for the variable x in an open statement such as $P(x) \wedge Q(x)$, the same element is substituted at each occurrence of the variable. Thus, with the set of natural numbers as universe of discourse, for example, the open statement $P(x) \wedge Q(x)$ would give rise to a closed statement such as $P(2) \wedge Q(2)$, but not to a closed statement such as $P(2) \wedge Q(3)$. This agrees with the use of variables in elementary algebra.

If $P(x_0)$ is a tautology for each element $x_0 \in U$, then $P(x)$ is said to be an open statement tautology. For example $P(x) \vee \frown P(x)$ is an open statement tautology for every open statement $P(x)$.

You may find the manipulation of these abstract notions easier on observing that you have used open statements frequently in the past. For example, the pair of points at which the parabola $y = x^2 - 4x$ intersects the x-axis is the set $\{x \mid x^2 - 4x = 0\}$, where it is understood from the context that an appropriate universe of discourse for x is the set R of real numbers. In this case the open statement is '$x^2 - 4x = 0$', and we naturally call such statements equations.

Given an open statement $P(x)$ with a universe of discourse U for the variable x, one is often concerned with the subset of values of x for which the corresponding closed statement $P(x_0)$ is true relative to the basic statements. Thus, in the above example it is understood that the basic statements are a sufficient set of basic statements for arithmetic, and R is taken as the universe of discourse of x. Then the set of values such that $x^2 - 4x = 0$ is true in the sense of ordinary arithmetic is the subset $\{0, 4\}$ of R. The subset of U comprising those elements x_0 for which $P(x_0)$ is a true statement is sometimes called the *truth set* of the open statement $P(x)$. It is precisely this set for which the notation $\{x \mid P(x)\}$ was introduced in §1.2. In the above example the set $\{0, 4\}$ is the truth set of the open statement $x^2 - 4x = 0$.

In elementary algebra, identities such as $(x^2 - 1) \equiv (x + 1)(x - 1)$ are particularly important. The identity sign is used to signify that the equation $(x^2 - 1) = (x + 1)(x - 1)$ is true in the sense of ordinary arithmetic for all real values of x. In other words, the truth set of the open statement $(x^2 - 1) = (x + 1)(x - 1)$ is the set R. In general, a case of particular importance is that in which the truth set for an open statement $P(x)$ is the whole of the universe of discourse U of the variable x, i.e. for each element x_0 of U the closed statement $P(x_0)$ is true relative to the basic statements. Thus one commonly considers such sentences as 'for all $x, P(x)$ is true', 'for each $x, P(x)$ is true', 'for every $x, P(x)$ is true'. These all mean the same thing, although we may use different wording on different occasions to achieve a change in emphasis. These statements are symbolized by

$$\forall x, P(x)$$

commonly read 'for all $x, P(x)$'. Notice that $\forall x, P(x)$ is a closed statement, since it is assigned truth value T or F according as the truth set of $P(x)$ is the whole of U or not. In particular, if the universe of discourse is a finite set, $U = \{a, b, c\}$ say, then the truth set of $P(x)$ is the whole of U precisely when $P(a) \wedge P(b) \wedge P(c)$ is true. Thus in this case we have the tautological equivalence

$$\forall x, P(x) \Leftrightarrow (P(a) \wedge P(b) \wedge P(c)).$$

In other circumstances one may only require to know that the truth set of an open statement is non-empty. In this connection statements like 'for at least one x, $P(x)$ is true', 'for some x, $P(x)$ is true', 'there exists x such that $P(x)$ is true' are used. These will all denote the same situation, symbolized by

$$\exists x, P(x)$$

and commonly read as 'there exists x such that $P(x)$'. Again $\exists x, P(x)$ is a closed statement, having truth value T whenever the truth set of $P(x)$ is non-empty, and having truth value F if the truth set of $P(x)$ is the empty set $\varnothing$. Thus if $U = \{a, b, c\}$, we have the tautological equivalence

$$\exists x, P(x) \Leftrightarrow (P(a) \vee P(b) \vee P(c)).$$

If the universe of discourse changes during a discussion, for example when a new open statement is introduced, then for clarity we shall fill out the statement $\exists x, P(x)$ to $\exists x \in U, P(x)$. The statement $\forall x, P(x)$ can be filled out to either $\forall x \in U, P(x)$ or $\forall x, x \in U \Rightarrow P(x)$, whichever is most helpful in the context. Similarly, the truth set of the open statement $P(x)$ can be denoted by either $\{x \in U \mid P(x)\}$ or $\{x \mid x \in U \text{ and } P(x)\}$.

The symbols $\forall$ and $\exists$, representing as they do phrases which give an indication of the number of elements x_0 for which $P(x_0)$ is true relative to the basic statements, are called quantifiers; $\forall$ is called the *universal quantifier* and $\exists$ is called the *existential quantifier*.

Closed statements obtained from open statements by means of quantifiers may themselves be either true or false relative to the basic statements from which they are constructed. Note that if $P(x)$ is an open statement tautology then the closed statement $\forall x, P(x)$ is true. We give now some numerical examples in which truth relative to the basic statements can be interpreted as truth in the ordinary arithmetic sense.

Example 1   Let $U = \{1, 2\}$ and let $P(x)$ denote '$x > 0$'. Then $P(1)$ is '$1 > 0$' which is true, while $P(2)$ is '$2 > 0$' which is also true. Hence the statement '$\forall x, P(x)$' is true. Also the statement '$\exists x, P(x)$' is true.

Example 2   Let $U = \{1, 3\}$ and let $P(x)$ denote '$x > 2$'. Then $P(1)$ is '$1 > 2$' which is false, while $P(3)$ is '$3 > 2$' which is true. Hence the statement '$\forall x, P(x)$' is false. However, the statement '$\exists x, P(x)$' is true.

To decide whether a quantified statement is true or false one often needs to write its negation in a convenient form. The negation of $\forall x, P(x)$ can certainly be written in the form $\sim[\forall x, P(x)]$. In order to determine an alternative form for the negation we examine first a situation in which the universe of discourse for the variable x has just three elements. If $U = \{a, b, c\}$, then $\forall x, P(x)$ is tautologically equivalent to $P(a) \wedge P(b) \wedge P(c)$. By de Morgan's law $\sim(P(a) \wedge P(b) \wedge P(c))$ is equivalent to $\sim P(a) \vee \sim P(b) \vee \sim P(c)$, which is equivalent to $\exists x, \sim P(x)$. The argument above, suitably extended, proves the equivalence of the two statements $\sim \forall x, P(x)$ and $\exists x, \sim P(x)$ whenever the universe of discourse for the variable x is a finite set. The argument cannot be extended to infinite universes of discourse because we have not admitted into our logical model statements involving an infinite collection of connectives. In order to decide whether to take as axiomatic in the model the equivalence of the two quantified statements we must pay regard to the usage of their analogues in everyday language, i.e. in the situation we are modelling. Here we say that if it is false that, for all possible values x_0 of the variable x, $P(x_0)$ is true relative to the basic statements, then we must be able to find at least one value, say x_1, for which $P(x_1)$ is false relative to the basic statements, i.e. for which the statement $\sim P(x_1)$ is true. The above discussion suggests that we do use these statements in the way indicated, and so in our model we shall take it as axiomatic that the two statements $\sim\forall x, P(x)$ and $\exists x, \sim P(x)$ are logically equivalent. Similar discussion leads us to regard the statement $\sim\exists x, P(x)$ as being equivalent to $\forall x, \sim P(x)$, the universe of discourse being the same for the variable x in each of the two statements.

We therefore have the two tautological equivalences

$$\sim\forall x, P(x) \Leftrightarrow \exists x, \sim P(x),$$

$$\sim\exists x, P(x) \Leftrightarrow \forall x, \sim P(x).$$

When the statement $P(x)$ is of the form of an implication $p(x) \Rightarrow q(x)$, these two equivalences combined with the equivalence

$$\sim(p \Rightarrow q) \Leftrightarrow (p \wedge \sim q)$$

discussed in the last section, lead to the tautological equivalences

$$(\sim\forall x, p(x) \Rightarrow q(x)) \Leftrightarrow (\exists x, p(x) \wedge \sim q(x)),$$

$$(\sim\exists x, p(x) \Rightarrow q(x)) \Leftrightarrow (\forall x, p(x) \wedge \sim q(x)).$$

Let us now determine the truth or falsehood of the statement 'for all real numbers x, the quadratic x^2+4x+3 is positive'. The statement can be expressed in the form

$$\forall x, x \in R \Rightarrow x^2+4x+3 > 0,$$

and its negation in the form

$$\exists x, x \in R \wedge \sim(x^2+4x+3 > 0),$$

i.e.

$$\exists x, x \in R \wedge x^2+4x+3 \leqslant 0.$$

The number $x = -2$ demonstrates that the last statement is true. Thus the original statement is false.

Exercises 2.5

1. Write each of the following statements using the symbols for

quantifiers and state which of them are true in ordinary arithmetic.

(a) Every integer is a real number.

(b) Every real number is rational.

(c) There exists a negative rational number.

(d) All the roots of the equation $x^2 - 5x - 6 = 0$ are positive.

(e) Every integer satisfies the equation $\cos \pi x = 1$.

(f) Every natural number is less than its square.

2. Express each of the following statements in English, and state which are true.

(a) $\forall x \in Z, (0 < x < 10) \wedge (x > 5)$.

(b) $\exists x \in N, (1 < x < 2) \vee (x = 4)$.

(c) $\forall x \in \{1, 2, 3\}, x^3 - 6x^2 + 11x - 6 = 0$.

(d) $\exists x \in \{5, 6, 7\}, x^2 - 8x + 5 > 0$.

(e) $\forall x \in R, x^2 + 2x + 1 = 0 \Rightarrow x^2 = 1$.

(f) $\forall x \in R, \cos x = 0 \Rightarrow \sim (x \in Q)$.

3. The following statements are true so long as in each case one chooses a universe of discourse from the sets of numbers N, Z, Q and R. In each case decide which sets may be taken as the universe of discourse in order to make the statement true.

(a) $\exists x, x^2 = 2$.

(b) $\forall x, x + 4 > 0$.

(c) $\exists x, 4x^2 = 1$.

(d) $\forall x, x^2 \in N$.

4. Discuss whether or not the following implications and the four reverse implications should be taken to be tautologies. You should consider finite universes of discourse first, and then everyday usage of statements in the required form, as was done in the discussion above concerning negation of quantified statements.

(a) $\forall x, p(x) \vee q(x) \Rightarrow \forall x, p(x) \vee \forall x, q(x)$.

(b) $\forall x, p(x) \wedge q(x) \Rightarrow \forall x, p(x) \wedge \forall x, q(x)$.

(c) $\exists x, p(x) \vee q(x) \Rightarrow \exists x, p(x) \vee \exists x, q(x)$.

(d) $\exists x, p(x) \wedge q(x) \Rightarrow \exists x, p(x) \wedge \exists x, q(x)$.

5. Write the definitions of $\bigcup_{i=1}^{n} A_i$ and $\bigcap_{i=1}^{n} A_i$ in rule form, expressing the rules as statements involving quantifiers. Discuss the extension of the notions of union and intersection to infinitely many sets.

6. Write the negations of the following statements in the alternative form developed above.

(a) $\forall x \in R, (x < 2) \vee (x > 2)$.

(c) $\exists x \in Q, (x^2 \in Z) \wedge \sim (x \in Z)$.

(b) $\exists x \in N, x^2 + x - 1 = 0$.

(d) $\forall x \in R, x^2 > 4 \Rightarrow x > 2$.

2.6 STATEMENTS INVOLVING TWO VARIABLES

So far we have encountered open statements involving just one variable, and we have used just one quantifier to form a closed statement. We shall now consider open statements involving two variables, such as '$x^2 - 2xy + 3y^2 = 0$'. Each variable has an associated universe of discourse, and each variable may be quantified.

Let us consider a general open statement $P(x, y)$, where U is the universe of discourse for the variable x, and V is the universe of discourse for the variable y. The variables x and y may be quantified in several ways and a discussion of one particular way will exemplify many important points.

We shall consider initially a finite example, with $U = \{a, b\}$ and $V = \{c, d\}$. The statement $\forall y \in V, P(x, y)$ is equivalent to the statement $P(x, c) \land P(x, d)$. Thus $\forall y \in V, P(x, y)$ is an *open* statement with the single variable x. The statement $\forall x \in U, \forall y \in V, P(x, y)$ is therefore equivalent to the *closed* statement

$$(P(a, c) \land P(a, d)) \land (P(b, c) \land P(b, d)).$$

Using the associative and commutative laws for conjunction, this can be seen to be equivalent to the statement

$$(P(a, c) \land P(b, c)) \land (P(a, d) \land P(b, d)),$$

which in turn is equivalent to the quantified statement

$$\forall y \in V, \forall x \in U, P(x, y).$$

We now have to extend the discussion to non-finite universes U, V. This will involve a detailed examination of the relationship between the open and closed statements arising, and their use in the situations we are trying to model.

As we saw above, the statement $\forall y \in V, P(x, y)$ is an open statement in the single variable x. It becomes a closed statement if we substitute a particular element x_0 of U for the variable x. The statement $\forall y \in V, P(x_0, y)$ is then true or false according as the truth set of the open statement $P(x_0, y)$—an open statement with the single variable y—is the whole of V or not.

Since $\forall y \in V, P(x, y)$ is an open statement in the single variable x, it may be quantified to yield the closed statement $\forall x \in U, \forall y \in V, P(x, y)$. The above discussion suggests that this should have truth value T if and only if for each particular element $x_0 \in U$ and for each particular $y_0 \in V$, the closed statement $P(x_0, y_0)$ has truth value T. Thus in ordinary usage there is symmetry between the two variables, and this, together with the finite example already considered, suggests that in the model the order of quantification should make no difference. We thus take as axiomatic the equivalence of the two statements $\forall x \in U, \forall y \in V, P(x, y)$ and $\forall y \in V, \forall x \in U, P(x, y)$. If $U = V$, this notation is commonly abbreviated to $\forall x, y \in U, P(x, y)$.

If the two variables in an open statement are quantified differently then the situation is altered, as an example from mathematical usage will show.

Consider the open statement $x < y$, where the universe of discourse for both x and y will be taken to be the set R of real numbers. The quantified statement $\forall x, \exists y, x < y$ corresponds to the statement that, given an arbitrary real number x_0 we can find an associated real number y_0 such that $x_0 < y_0$, and this is true relative to the basic statements of arithmetic. The quantified statement $\exists y, \forall x, x < y$ corresponds to the statement that there is a real number y_0 greater than every real number, and this is false in arithmetic. Thus in our model we must treat the two statements $\forall x \in U, \exists y \in V, P(x, y)$ and $\exists y \in V, \forall x \in U, P(x, y)$ as being different, and care must be taken not to confuse the two.

We have not considered all the possibilities concerning two variables, others are explored in the exercises at the end of this section. Similar problems arise with statements involving more than two variables.

The translation between statements in everyday mathematics and their analogues as quantified statements in logic is a process which entails some difficulties, especially when more than one quantifier is involved, and so we give two examples.

Consider the statement 'there exists a least natural number'. This statement asserts the existence of a particular natural number with special properties, and so the corresponding quantified statement should begin '$\exists x \in N$', followed by a formulation of the property x is to have. This property must be formulated as an open statement involving the variable x, and symbols from arithmetic. A verbal statement expressing the property in question is 'x is a least natural number'. It must now be recognized that the notion of 'least' is connected with the arithmetical concept of inequality. Thus 'x is a least natural number' can be re-phrased as 'x is less than every other natural number' or 'x is less than or equal to every natural number'. In this form the statement is seen to relate x to every natural number y, and so it can be formulated as '$\exists x \in N, \forall y \in N, x \leqslant y$'.

As an example of the converse process we consider the logical statement

$$\forall x \in N, \exists y \in R, (y > 0) \wedge (y^2 = x).$$

A literal translation of this would be 'for every natural number x there exists a real number y such that y is positive and $y^2 = x$'. We can now use arithmetical terminology to call y the square root of x, and we can then drop the variables and condense the statement to 'every natural number has a positive real square root'.

Exercises 2.6

1. Write each of the following statements using the symbols for quantifiers and state which of them are true in ordinary arithmetic.
   (a) There exists an odd number which is an integer power of two.
   (b) The square of every integer has remainder 0 or 1 on division by 4.
   (c) Corresponding to each irrational number x, there is an integer y satisfying $x < y < x+1$.
   (d) Every integer can be expressed as a sum of two perfect squares.

2. Express the following statements in English, and state which are true.
   (a) $\forall x \in R, \exists y \in R, x^2 - 5xy + 6y^2 = 0$.
   (b) $\exists y \in R, \forall x \in R, x^2 - 5xy + 6y^2 = 0$.
   (c) $\forall x \in R, \forall y \in R, (x < y \wedge y \neq 0) \Rightarrow x/y < 1$.
   (d) $\forall x \in Z, \exists y \in Z, x = 2y \vee x = 2y+1$.

3. Discuss the meaning of each of the two statements

$$\exists x \in U, \exists y \in V, P(x, y), \qquad \exists y \in V, \exists x \in U, P(x, y),$$

and give reasons why it is reasonable to postulate their equivalence. (If $U = V$ we abbreviate the above statements to $\exists x, y \in U, P(x, y)$.)

4. Suppose that the statement $\exists y \in V, \forall x \in U, P(x, y)$ has truth value T. Decide whether the statement $\forall x \in U, \exists y \in V, P(x, y)$ should be postulated as having truth value T.

5. Write the negations of the following statements in the alternative form developed in §2.5.

    (a) $\exists x, \forall y, x < y.$ (c) $\forall x, \forall y, \exists z, xy = z^2.$

    (b) $\forall x, \exists y, \forall z, xy < z.$

6. Let the universe of discourse be U. Discuss the difference between the statements

    (a) $\forall A \subseteq U, \exists X \subseteq U, A \cap X = A,$

    (b) $\exists X \subseteq U, \forall A \subseteq U, A \cap X = A,$

and between the statements

    (c) $\forall A \subseteq U, \exists X \subseteq U, A \cup X = A,$

    (d) $\exists X \subseteq U, \forall A \subseteq U, A \cup X = A.$

Compare these statements with the statements

    (e) $\forall a \in R, \exists x \in R, a + x = a,$ (g) $\forall a \in R, \exists x \in R, a.x = a,$

    (f) $\exists x \in R, \forall a \in R, a + x = a,$ (h) $\exists x \in R, \forall a \in R, a.x = a.$

7. Discuss the existence of solutions of the equations $A \cup X = U$ and $A \cap X = \emptyset$.

2.7 MATHEMATICAL PROOF

It is not the aim of mathematics to declare what is true and what is false in the world, but rather to produce conceptual models of certain aspects of the world and to study within the models which statements are logical consequences of other statements. For example, Euclidean geometry was designed as a model for surveying and it consists of the study of those theorems which follow from certain statements about lines and points, including a statement about parallel lines. In order to understand more fully the role of the Euclidean assumption about parallel lines, mathematicians have studied non-Euclidean geometries in which the assumption about parallel lines is changed. Although many of the mathematical statements which you have studied have the appearance of absolute truth, they depend on the assumption of a number of generally acceptable basic statements. The basic statements for a branch of mathematics are called the *axioms* for that branch. A statement in a branch of mathematics is said to be true if it is true relative to the axioms, i.e. if when the axioms are given the truth value T the statement also has the truth value T. We are concerned here with the way in which statements are shown to be true on the basis of a set of axioms. We recall from §2.4 that the compound statement P is true relative to a set of axioms $a_1, a_2, \ldots, a_n$ if and only if the statement $a_1 \wedge a_2 \wedge \ldots \wedge a_n \Rightarrow P$ is a tautology. However, because we study complicated statements P, it is rare that such a statement can be shown to be true by expressing it in terms of the axioms and logical connectives, and by using a single truth table to prove the existence of the required tautology. Most statements are expressed in terms of subsidiary concepts and symbols, and the existence of the required tautology is demonstrated by the use of a sequence of simpler tautologies.

A definition of a subsidiary concept or symbol plays the same part as a tautological equivalence in the sense that the role of a definition is to replace one phrase by another, this being generally done for reasons of economy of expression. For example, a circle may be defined by the statement 'a circle is a set of points in a plane equidistant from a fixed point'. What is done here is to say that whenever we encounter a set of points in a plane equidistant from a fixed point, we shall describe the set by using the word 'circle'. Another function of a definition is to express a statement, which may be an intuitive one, in terms of basic statements, and thus render the original statement more susceptible to logical analysis. This is done for example in the study of the notions of limit and continuity. Definitions in everyday mathematical language are often phrased as in the statement above concerning the circle, but in the present section we shall use 'P for Q' to denote that P is defined by Q, where P and Q may be statements, or descriptions of mathematical objects. This will emphasize that, logically, P and Q mean exactly the same. Naturally their psychological impacts may be very different.

Formally, a mathematical proof will consist of a sequence of true statements. A statement may be true because it is a particular case of a tautology; it may be taken to be true because it is one of the basic statements or axioms, or comes from a definition; or it may be inferred to be true because it is tautologically equivalent to or tautologically implied by one or more of the previous true statements.

A mathematical proof may be presented formally with the steps enumerated for easy reference. However, this is a cumbersome procedure even for the simplest propositions, and proofs are usually presented in a discursive form in which the main steps of a formal proof are indicated. The author sets out as many of the steps as he thinks necessary to convince him that both he and the prospective reader can fill out all the details. Thus the proof of a proposition which takes a page of an elementary text may be dismissed in two lines in a more advanced book. We aim at proofs presented in the discursive style, and only present the more formal style of proof for a few propositions in order to show how the proofs are based on the tautologies of §2.4.

We shall now prove some of the results of §1.6 by the methods outlined above. The basic statements concern the relationships between the symbols we used when discussing sets, and they are presented using the quantifiers introduced in §2.5.

Basic statements

$$\text{I} \quad A \subseteq B \text{ for } \forall x, x \in A \Rightarrow x \in B.$$
$$\text{II} \quad A = B \text{ for } (A \subseteq B) \wedge (B \subseteq A).$$
$$\text{III} \quad x \in A \cup B \text{ for } (x \in A) \vee (x \in B).$$
$$\text{IV} \quad x \in A \cap B \text{ for } (x \in A) \wedge (x \in B).$$

These basic statements are only concerned with the way in which the language and notations of sets are used: they are not axioms for formal set theory.

As a first example we take a statement which is shown to be equivalent to one of the standard tautologies, and which is therefore true relative to the basic statements. Two presentations of the proof are given. In the first we display formally the translation of the statement $A \cap B \subseteq A$ into a statement involving $x \in A$ and $x \in B$ which is identified as a tautology. In this case we give the formal proof in detail, but in later examples we shall exercise a certain amount of licence in the interests of clarity. The second presentation of the proof is the usual kind of argument which is employed in mathematics. We then discuss the relationship between the two proofs.

Proposition 2.1 Let A and B be any two sets. Then $A \cap B \subseteq A$.

First presentation of proof

1. $A \cap B \subseteq A$ for $(\forall x, x \in A \cap B \Rightarrow x \in A)$ (I)
2. $(\forall x, x \in A \cap B \Rightarrow x \in A)$ for $(\forall x, (x \in A \wedge x \in B) \Rightarrow x \in A)$ (IV)

E

3. $A \cap B \subseteq A$ for $(\forall x, (x \in A \land x \in B) \Rightarrow x \in A)$ (from 1 and 2)

4. $\forall x, (x \in A \land x \in B) \Rightarrow x \in A$ (law of simplification, tautology vii)

5. $A \cap B \subseteq A$ (from 3 and 4)

Second presentation of proof   Let x_0 be an arbitrary element of $A \cap B$. Then $x_0 \in A$ and $x_0 \in B$, by definition. Thus, in particular, $x_0 \in A$. But x_0 was arbitrary, and so $A \cap B \subseteq A$, by definition.

Comments   In the first presentation of the proof, steps 1 to 3 consist of a translation of the proposition into a quantified statement with connectives. In step 4 we recognize that the quantified statement is a true closed statement which arises from an open statement tautology, and in step 5 we infer from this the desired result.

In the second presentation of the proof it is assumed that the reader has translated the proposition into a quantified statement with connectives. In that proof the relationship between the open statement $x \in A \cap B \Rightarrow x \in A$ and the closed statement $x_0 \in A \cap B \Rightarrow x_0 \in A$ is indicated by the use of the word 'arbitrary'. It is meant to convey the idea that an argument which is valid for an element x_0 chosen at random will be valid for any other particular element, so that the truth value of the quantified statement $\forall x, x \in A \cap B \Rightarrow x \in A$ will be the same as the truth value of the closed statement $x_0 \in A \cap B \Rightarrow x_0 \in A$.

The explicit translation of $A \cap B \subseteq A$ which was completed at step 3 of the first proof has been omitted in the second proof. This will not always be the case; indeed, even in informal proofs often the most important part of the whole procedure is the transformation of the required statement into a form which is technically easier to handle.

The proof that $A \cap B \subseteq A$ is valid for every pair of sets A and B. Thus we could express the proposition in the form $\forall A, \forall B, A \cap B \subseteq A$. However, if the proposition were expressed in this way, the quantifications $\forall A, \forall B$ would have to be used throughout the first presentation of the proof, and the second proof would have to commence with the sentence 'Let A_0 and B_0 be any two sets.' This would needlessly complicate matters and would be unlikely to increase the understanding of the processes of proof, so we shall not quantify

A and *B* when these symbols appear in the propositions of this section.

For most propositions the expanded forms of the statements of the type given in the first proof of Proposition 2.1 are too long to be handled conveniently. Even for the simple statement $A \cup B = B \cup A$ it is convenient to separate out parts of the statement and put them together again later.

Proposition 2.2   Let *A* and *B* be any two sets. Then $A \cup B = B \cup A$.

First presentation of proof

1. $A \cup B = B \cup A$ for $(A \cup B \subseteq B \cup A) \wedge (B \cup A \subseteq A \cup B)$ (II)
2. $A \cup B \subseteq B \cup A$ for $(\forall x, x \in A \cup B \Rightarrow x \in B \cup A)$ (I)
3. $A \cup B \subseteq B \cup A$ for $(\forall x, (x \in A \vee x \in B) \Rightarrow (x \in B \vee x \in A))$ (III)
4. $\forall x, (x \in A \vee x \in B) \Rightarrow (x \in B \vee x \in A)$
   (commutative law for disjunction, tautology i)
5. $A \cup B \subseteq B \cup A$ (from 3 and 4)
6. $B \cup A \subseteq A \cup B$ for $(\forall x, x \in B \cup A \Rightarrow x \in A \cup B)$ (I)
7. $B \cup A \subseteq A \cup B$ for $(\forall x, (x \in B \vee x \in A) \Rightarrow (x \in A \vee x \in B))$ (III)
8. $\forall x, (x \in B \vee x \in A) \Rightarrow (x \in A \vee x \in B)$
   (commutative law for disjunction, tautology i)
9. $B \cup A \subseteq A \cup B$ (from 7 and 8)
10. $A \cup B = B \cup A$ (from 1, 5 and 9)

Second presentation of proof   Let x_0 be an arbitrary element of $A \cup B$. Then by definition $x_0 \in A$ or $x_0 \in B$. Thus $x_0 \in B$ or $x_0 \in A$, i.e. $x_0 \in B \cup A$. Since x_0 was arbitrary, $A \cup B \subseteq B \cup A$. Similarly $B \cup A \subseteq A \cup B$. Hence by definition $A \cup B = B \cup A$.

Comments   In the first proof, statements 4 and 8 involve a tautology of the form $p \vee q \Rightarrow q \vee p$. To establish this from the commutative law for disjunction, which states that $p \vee q \Leftrightarrow q \vee p$, we can express $\Leftrightarrow$ in terms of $\Rightarrow$ and $\wedge$, as was done in the initial discussion of § 2.4, and then use the law of simplification.

In the second presentation of the proof the fact that we have used the commutative law for disjunction has not been explicitly stated, and similarly the full statement of relevant definitions is not included; it is assumed that readers know or can find these, and can fill out the logical background to the proof. The repetition of the form of argument with no more than the interchange of symbols which occurred in steps 6–9 in the first proof is avoided in the second by the use of the word 'similarly'.

We next illustrate a method of proof of a conditional statement. The meaning of such a statement in terms of truth values is that if the axioms are given the truth value T and if the condition is given the truth value T, then the conclusion also has the truth value T. In this example the conclusion is the conjunction of $A \cap B \subseteq A$ and $A \subseteq A \cap B$. The first of these has already been proved true relative to the basic statements, and this result will be used in the following proof.

Proposition 2.3   If $A \subseteq B$ then $A \cap B = A$.

First presentation of proof

1. $A \subseteq B$ for $(\forall x, x \in A \Rightarrow x \in B)$ (I)
2. $(\forall x, x \in A \Rightarrow x \in B) \Rightarrow (\forall x, x \in A \Rightarrow (x \in A \wedge x \in B))$
                                                    (tautology xi)
3. $A \subseteq A \cap B$ for $(\forall x, x \in A \Rightarrow (x \in A \wedge x \in B))$ (I and IV)
4. $A \subseteq B \Rightarrow A \subseteq A \cap B$ (from 1, 2 and 3)
5. $A \cap B \subseteq A$ (proposition 2.1)
6. $A \subseteq B \Rightarrow A \cap B \subseteq A$ (from 5, using tautology vi)
7. $A \subseteq B \Rightarrow (A \subseteq A \cap B \wedge A \cap B \subseteq A)$
                                     (from 4, 6 using tautology x)
8. $A \subseteq B \Rightarrow A = A \cap B$ (II)

Second presentation of proof   Let x_0 be an arbitrary element of A. Since $A \subseteq B$, $x_0 \in B$ so that $x_0 \in A$ and $x_0 \in B$. Since x_0 was arbitrary it follows that $A \subseteq A \cap B$. Moreover, by proposition 2.1, $A \cap B \subseteq A$. Hence $A \cap B = A$.

Comments   The most significant difference between the proofs is that in the first we have proved directly that the given statement is true relative to the basic statements, whereas in the second proof we have assumed the condition $A \subseteq B$ to be true, and from this inferred that the conclusion is also true. This fits in with the logical structure of the first proof due to the fact that if the condition has truth value F then the statement is true irrespective of the truth value of the conclusion. This procedure of assuming the truth of the condition and inferring the truth of the conclusion will be followed in most mathematical proofs; indeed this is generally the accepted meaning of a conditional statement in ordinary usage.

We now turn to the empty set $\emptyset$. We found in §1.3 that verbal arguments in the style given there were not convincing so that here the formal method will carry the burden of proof. Because of this, we use the notation $\sim x \in A$ rather than the notation $x \notin A$ for the negation of the statement $x \in A$. First we apply formal methods to prove the following proposition.

Proposition 2.4 If $\forall x, \sim x \in A$ and $\forall x, \sim x \in B$ then $A = B$.

Proof

1. $(\forall x, \sim x \in A \land \forall x, \sim x \in B) \Rightarrow (\forall x, \sim x \in A)$ (tautology vii)
2. $(\forall x, \sim x \in A) \Rightarrow (\forall x, \sim x \in A \lor x \in B)$ (tautology vi)
3. $(\forall x, \sim x \in A \lor x \in B) \Rightarrow (\forall x, \ x \in A \Rightarrow x \in B)$ (tautology xiii)
4. $(\forall x, \sim x \in A \land \forall x, \sim x \in B) \Rightarrow (\forall x, x \in A \Rightarrow x \in B)$
       (from 1, 2, 3 using transitivity)
5. $(\forall x, \sim x \in A \land \forall x, \sim x \in B) \Rightarrow A \subseteq B$ (from 4, using I)
6. $(\forall x, \sim x \in A \land \forall x, \sim x \in B) \Rightarrow B \subseteq A$ (by a similar argument)
7. $(\forall x, \sim x \in A \land \forall x, \sim x \in B) \Rightarrow (A \subseteq B \land B \subseteq A)$
       (from 5, 6 using tautology x)
8. $(\forall x, \sim x \in A \land \forall x, \sim x \in B) \Rightarrow A = B$ (from 7, using II)

Comments The proof above shows that there is at most one set with the property of having no elements, it does not demonstrate the existence of such a set.

We now agree to accept in the formal system a set $\emptyset$, called the empty set, whose use is specified by a fifth basic statement.

Basic Statement V $\forall x, \sim x \in \emptyset$.

The statement that the empty set is a subset of every set was discussed informally in §1.4. A formal proof of the statement is given here.

Proposition 2.5 Let A be any set. Then $\varnothing \subseteq A$.

Proof

1. $\varnothing \subseteq A$ for $(\forall x, x \in \varnothing \Rightarrow x \in A)$ (I)
2. $\forall x, \ x \in \varnothing \Rightarrow x \in A) \Leftrightarrow (\forall x, \sim x \in \varnothing \vee x \in A)$ (tautology xiii)
3. $(\forall x, \sim x \in \varnothing) \Rightarrow (\forall x, \sim x \in \varnothing \vee x \in A)$ (tautology vi)
4. $\forall x, \sim x \in \varnothing \Rightarrow \varnothing \subseteq A$ (from 1, 2 and 3, using transitivity)
5. $\forall x, \sim x \in \varnothing$ (V)
6. $\varnothing \subseteq A$ (from 4 and 5)

We have tried to develop some ideas of mathematical proof, and to relate them to the procedure of logical inference from a set of basic statements. The basic statements were made quite explicit in the examples above, but often the dependence on basic statements is not so apparent, particularly when an extensive theory has been constructed with numerous subsidiary concepts and propositions. Indeed the task in many investigations is to discover a set of basic statements or axioms which can serve as a foundation for an intuitively based theory. One of the important developments of calculus during the nineteenth century was a search of this kind.

In an extensive theory, the task of proving later propositions by direct proofs of the kind illustrated above may be more difficult than the task of inferring the truth of the propositions by indirect means. In the examples which follow we give instances of indirect proofs in situations where the basic statements are not made explicit. We give only discursive proofs, but we discuss the logical structure of these proofs.

To explore indirect proofs we first look at various combinations of implication and negation. Table 2.14 gives the truth tables of the statements $p \Rightarrow q$, $\sim q \Rightarrow \sim p$, $q \Rightarrow p$, $\sim p \Rightarrow \sim q$.

If $p \Rightarrow q$ is the original statement, $\sim q \Rightarrow \sim p$ is called the *contrapositive*. That a statement and its contrapositive are tautologically equivalent is easily deduced from Table 2.14. The tautology is commonly called the *law of contrapositive*, as was indicated in the

TABLE 2.14

p	q	$\sim p$	$\sim q$	$p \Rightarrow q$	$\sim q \Rightarrow \sim p$	$q \Rightarrow p$	$\sim p \Rightarrow \sim q$
T	T	F	F	T	T	T	T
T	F	F	T	F	F	T	T
F	T	T	F	T	T	F	F
F	F	T	T	T	T	T	T

list at the end of §2.4. The statement $q \Rightarrow p$ is called the *converse* of $p \Rightarrow q$, and $\sim p \Rightarrow \sim q$ is called the *inverse* of $p \Rightarrow q$. The table shows that the converse is tautologically equivalent to the inverse. Table 2.14 also shows that the statement $p \Rightarrow q$ and its converse $q \Rightarrow p$ are not logically equivalent, so that in some circumstances $p \Rightarrow q$ may be true relative to a set of basic statements whilst $q \Rightarrow p$ is false relative to those basic statements. We illustrate this with an arithmetic example.

Let p stand for the statement '$m.n$ is an even integer'. Let q stand for the statement 'm is an even integer' and let r stand for the statement 'n is an even integer'. Then the statement $(q \wedge r) \Rightarrow p$ is a true statement, for the product of two even integers is even. Its contrapositive is $\sim p \Rightarrow \sim (q \wedge r)$, which is equivalent to $\sim p \Rightarrow (\sim q \vee \sim r)$ by de Morgan's law. The contrapositive says that if $m.n$ is odd then m is odd or n is odd. The converse statement is $p \Rightarrow (q \wedge r)$, which says that if $m.n$ is even then both m and n must be even, and this is false. Thus we have an example of a true conditional statement whose converse is false.

Because of the fact that a conditional statement and its contrapositive are equivalent we can use the contrapositive as a basis for an indirect proof, as in the following proposition, in which the universe of discourse for the variable a is the set of integers.

Proposition 2.6 If a^2 is an even integer then a is an even integer.

Discussion of proof If p stands for the statement 'a^2 is an even integer' and q stands for the statement 'a is an even integer' then we wish to prove that $p \Rightarrow q$ is a true statement. In fact we prove that $\sim q \Rightarrow \sim p$ is a true statement. Stated in words this says that if a is an odd integer then a^2 is an odd integer. The usual form of presentation of such a proof is as follows.

Proof Suppose a is an odd integer. Then $a = 2n+1$, where n is an integer. Hence $a^2 = 4n^2+4n+1 = 2(2n^2+2n)+1$. Thus a^2 is an odd integer. Hence if a^2 is even then a is even.

Comments Implicit in the proof are many properties of integers which would be basic statements or consequences thereof. We have not stated explicitly in this short proof that we have used the law of contrapositive. However, we would make explicit mention of the law of contrapositive at a step in a long involved proof, and you may find it helpful to do so even here.

Another indirect proof is known as *reductio ad absurdum*—hence the name of tautology (xviii). In everyday language the procedure is as follows. Our assumption is p and we wish to prove q. We assume $\sim q$ as well as p and deduce a statement of the form $r \wedge \sim r$, which is a self contradiction. This then shows that if p is true then in fact $\sim q$ must be false, i.e. q is true.

Proposition 2.7 The number $\sqrt{2}$ is irrational.

Discussion of proof The proof depends on a translation of the proposition into a form which can be shown to be true by means of tautologies. By means of the introduction of a variable x, we translate the proposition into the statement '$x = \sqrt{2}$ implies that x is irrational'. Now $x = \sqrt{2}$ stands for $x^2 = 2$ and $x > 0$; and x is irrational stands for $\sim x \in Q$. In the manipulation of the proof we use the fact that rational numbers can be expressed by fractions in lowest terms. The relationship of the proof with the standard presentation of the tautology *reductio ad absurdum* is given by the correspondence—p for $x^2 = 2$; q for x is irrational; r for b/c is a fraction in lowest terms.

Proof Suppose $x^2 = 2$ and x is rational. Then $x = b/c$ where b/c is a fraction in lowest terms. Thus

$$b^2/c^2 = 2, \quad \text{i.e. } b^2 = 2c^2.$$

It follows that b^2 is even and so b is even by Proposition 2.6. Setting $b = 2m$, we see that $2m^2 = c^2$. Thus c^2 is even and so c is even. Hence b and c are both even, contradicting the fact that b/c is in its lowest terms. Thus the assumption that x is rational leads to a contradiction, and so x must be irrational.

We now give an example of a commonly encountered invalid deduction. Suppose that we know that $p \Rightarrow q$ is true and that q is true. Can we then deduce the truth of p? We have to investigate whether the statement $((p \Rightarrow q) \land q) \Rightarrow p$ is a tautology.

TABLE 2.15

p	q	$p \Rightarrow q$	$(p \Rightarrow q) \land q$	$((p \Rightarrow q) \land q) \Rightarrow p$
T	T	T	T	T
T	F	F	F	T
F	T	T	T	F
F	F	T	F	T

The statement is not a tautology and so the argument is not a valid one. We can easily construct an example of a situation where this kind of argument does lead to a false conclusion, as follows.

$-1 = 1$ implies $(-1)^2 = 1^2$, i.e. $1 = 1$. This latter statement is certainly true, and the implication is a valid one. However, the initial statement $-1 = 1$ is false.

We have been considering above the proving of statements, but sometimes we wish to disprove a statement p. In principle all we need do is to consider the statement $\sim p$ and prove that this is true. A particular case is when we wish to disprove a statement of the form $\forall x, P(x)$. We saw in §2.5 that the statement $\sim \forall x, P(x)$ is equivalent to the statement $\exists x, \sim P(x)$. Thus to show that $\forall x, P(x)$ is false we merely have to produce a single element x_0 from the universe of discourse such that the closed statement $P(x_0)$ is false. Such an element constitutes a *counterexample*. This process was demonstrated in §2.5 with the example 'for all real numbers x, the quadratic $x^2 + 4x + 3$ is positive'. The number $x = -2$ considered there is a counterexample.

In exploring the ideas and procedures involved in mathematical proof in such detail, we have tried to highlight some of the philosophical issues involved. Generally, mathematical activity consists of *doing* mathematics and *proving* statements, whereas here we have been concerned with talking *about* mathematics and talking *about* what constitutes proof. One way in which we hope such discussion can improve the understanding of mathematics is by enabling the logical structure of a proof to be split off from the manipulative details, and to this end we have endeavoured to keep the content of our examples as simple as possible. The most important thing which we hope that this chapter can accomplish is that it will promote thought and discussion about the issues involved, for there is no proof without *people*.

In future, you will have to decide on notations, on subsidiary concepts and symbols, and on the style in which you will present your proofs. You will have to decide how much of an argument to set down in order to convince yourself that there is no gap in your argument. You will also have to decide how much of an argument you need to set down in order to lead a prospective reader through your proof.

Exercises 2.7

1. Prove that the following statements are true for all sets A, B and C.
   (a) $A \subseteq A \cup B$.
   (b) $(A \cup B) \cup C = A \cup (B \cup C)$.
   (c) $A \cup (B \cap C) = (A \cup B) \cap (A \cup C)$.
   (d) $A \triangle B \subseteq A \cup B$.
   (e) $A = (A \backslash B) \cup (A \cap B)$.
   (f) $(A \backslash B) \cap (A \cap B) = \emptyset$.
   (g) $A \cup (B \backslash A) = A \cup B$.
   (h) $A \cap (B \backslash A) = \emptyset$.

2. Prove that the following statements are true for all sets A, B and C.
   (a) If $A \cap B = \emptyset$ then $A \backslash B = A$.
   (b) If $A \subseteq B$ then $A \backslash B = \emptyset$.
   (c) If $A \subseteq B$ then $C \cap A \subseteq C \cap B$.
   (d) If $A^c \subseteq B^c$ then $B \subseteq A$.
   (e) If $A \subseteq B$ and $B \subseteq C$ then $A \subseteq C$.

3. Write out the contrapositive, inverse and converse statements to the statements of exercise 2 and decide whether the converse statements are true.

4. Prove that if x is rational and y is rational, then $x + y$ is rational. Prove that if x rational and y is irrational, then $x + y$ is irrational.

5. Express the statement '$(\sqrt{2})^{-1}$ is irrational' as an implication $p \Rightarrow q$ and prove that it is true. Are the contrapositive, inverse and converse of your statement $p \Rightarrow q$ true?

6. Prove that the equation $ax^2 + bx + c = 0$ has two distinct real

roots if $a \neq 0$ and $b^2 - 4ac > 0$. Display the logical structure of your proof and discuss the basic assumptions about real numbers which you use. Is the converse of the statement true?

7. Write the following statement using the symbols for quantifiers. 'A real number which is less than every positive real number is not positive.' Prove the statement, displaying the logical structure of your proof and the basic assumptions about real numbers which you have used.

8. Discuss possible methods of proof of statements of the form $p \Rightarrow (q \vee r)$.

Prove that if $y > x^2 - 4$, then $y \geq 0$ or $|x| < 2$.

3

RELATIONS AND FUNCTIONS

3.1 INTRODUCTION

In Chapter 1 we noticed that in dealing with collections of objects, the order in which the objects in a given collection are described may or may not be important. The language of sets developed in Chapter 1 forms a model for those situations in which order is not important. In this chapter some situations where order is important are considered. We shall describe a conceptual model for situations involving statements about pairs of elements and relationships between elements.

Words which describe relationships of kinship are commonly used in statements about isolated pairs such as 'John is Jack's father'. This statement might convey different information to different people. To Jack's friend it might serve to introduce John, whereas to a stranger at a large family gathering it might help to sort out the generations and branches of the family. The friend and the stranger are likely to be interested in different aspects of the father–son relationship. If he was investigating the family tree, the stranger would only be concerned with the fact that two names in the tree are connected by this relationship. He would then regard the statement 'John is Jack's father' as a particular example of the general statement 'this one is that one's father'. He could note all the information he needed by listing pairs of names, father and son. Furthermore, if he wrote these pairs with the father's name first, then a list of such pairs would completely describe that aspect of the relationship in which he was interested.

There are many similar situations in which lists of pairs of elements can be used to describe some aspects of a problem. For example, if a man wished to fly from Edinburgh to Miami Beach he would first need to know whether it was at all possible, and then he would concern himself with choices of routes, of airlines, and of times. For the basic problem he would find useful a list of pairs of towns such that there are scheduled flights from the first town to the second. From this list he could pick out particular connections such as

Edinburgh to London, London to New York, and New York to Miami Beach.

In mathematics, the relationship of inequality gives rise to closed statements about particular pairs of numbers such as $2 < 4$. It also gives rise to open statements such as $x < 4, 2 < y$ and $x < y$. It is interesting to note how these three general statements focus one's attention on different aspects of the particular statement $2 < 4$. For many purposes all that one needs to know about the relationship of inequality are the truth sets of open statements of the type illustrated here. The truth sets of the first two open statements are sets of numbers. However, since the open statement $x < y$ involves two variables, its truth set will consist of pairs of numbers.

In many other situations, both in everyday life and in mathematics, the information can be transmitted as a collection of pairs. In all the examples above the elements of a pair have been of the same kind, but this need not be so. For example, one might consider a pair of elements, the first of which is a quadratic equation, and the second its set of solutions.

3.2 ORDERED PAIRS AND CARTESIAN PRODUCTS

The relationship discussed in §3.1 can be symbolized by means of pairs. However, the symbolism and language of sets does not easily lend itself to the requirement of indicating order or precedence. If we wish to symbolize the statement '$2 < 4$' considered as a special case of the open statement '$x < y$', then the symbol $\{2, 4\}$ is inadequate. Set notation does not distinguish between $\{2, 4\}$ and $\{4, 2\}$, whereas '$2 < 4$' and '$4 < 2$' are distinct statements, and indeed have different truth values. The required precedence can be indicated by using symbols from set theory (see exercise 5 below), but it is cumbersome and not easy to handle practically. We therefore introduce a new concept, that of an *ordered pair*, and we symbolize it by the notation (x, y) where the braces of set theory have been replaced by parentheses. Two ordered pairs (x_1, y_1) and (x_2, y_2) are said to be equal if and only if $x_1 = x_2$ and $y_1 = y_2$. Thus if $x \neq y, (x, y)$ and (y, x) are different. If we use the notation of ordered pairs in connection with the relationship of fatherhood discussed in §3.1, and agree that the father's name is to be the first component of the ordered pair, we may write (John, Jack), and this will carry the information that John is Jack's father. The concept will also distinguish between (2, 4) and (4, 2), so that ordered pairs can be used in connection with the statement '$x < y$'.

The notation of ordered pairs is used in coordinate geometry to distinguish the points (1, 2) and (2, 1) in the plane, and there are many other parts of mathematics in which ordered pairs of elements are used. For example, when you draw the graph of $y = \cos x$, you will make sure that it passes through the point (0, 1), and you will not confuse this with the point (1, 0). Naturally the inclusion of the point (0, 1) as a point of the graph is dependent upon the specification of the first component of the ordered pair as the value of the variable x, and of the second as the value of the variable y. In coordinate geometry one depicts this by using axes and by interpreting the components of the ordered pairs as distances relative to these axes. In the notation of sets, a graph such as the one discussed above could be written as the set of ordered pairs

$$\{(x, y) \mid x \in R \text{ and } y \in R \text{ and } y = \cos x\}.$$

In coordinate geometry the set of all ordered pairs whose components are real numbers is used as a universe of discourse. In other situations the set of all ordered pairs whose first components belong to a set A and whose second components belong to a set B may be used. This set of ordered pairs is called the *Cartesian product* of A with B, and is denoted by $A \times B$, which is read 'A cross B'. In the rule method of describing sets we have

$$A \times B = \{(x, y) \mid x \in A \text{ and } y \in B\}.$$

If A and B are small enough sets, then the elements of $A \times B$ can be listed. For example, if $A = \{1, 2, 3\}$ and $B = \{5, 6\}$ then

$$A \times B = \{(1, 5), (2, 5), (3, 5), (1, 6), (2, 6), (3, 6)\},$$
$$B \times A = \{(5, 1), (5, 2), (5, 3), (6, 1), (6, 2), (6, 3)\}.$$

This example shows that if A and B are different then $A \times B$ and $B \times A$ need not be equal. On the other hand $A \times A$ can be abbreviated to A^2 since no consideration of the order of the product is necessary. In this notation, the set used in coordinate geometry is the coordinate plane $R \times R = R^2$.

Exercises 3.2

1. Make sketches of the following subsets of the plane.

   (a) $T_1 = \{(x,y) \mid x = 3, y = 2\}$.
   (b) $T_2 = \{(x,y) \mid x = 2\}$.
   (c) $T_3 = \{(x,y) \mid y = x^4\}$.
   (d) $T_4 = \{(x,y) \mid x \leqslant y\}$.
   (e) $T_5 = \{(x,y) \mid x \leqslant y \text{ and } 0 \leqslant x \leqslant 1\}$.
   (f) $T_6 = \{(x,y) \mid x \leqslant y \text{ or } 0 \leqslant x \leqslant 1\}$.
   (g) $T_7 = \{(x,y) \mid x^2 + y^2 \geqslant 3\}$.
   (h) $T_8 = \{(x,y) \mid y > x^3 + 1\}$.

2. Let $A = \{a,b\}$ and $B = \{c,d,e\}$. Determine $A \times A$, $A \times B$, $B \times A$ and $B \times B$.

3. Let $A = \{1,2,3\}$. Make a sketch of the set $A \times A$ as a subject of the plane.

4. If A has m elements and B has n elements, where m and n are natural numbers, how many elements has $A \times B$?

5. Let $A = \{\{x\}, \{x,y\}\}$ and let $B = \{\{u\}, \{u,v\}\}$. Prove that $A = B$ if and only if $x = u$ and $y = v$.

6. Let A and B be non-empty sets. Prove that $A \times B = B \times A$ if and only if $A = B$.

7. Let $A = \{a_1, a_2\}$, $B = \{b_1, b_2, b_3\}$ and $C = \{c_1, c_2\}$. Determine $A \times B$, $B \times C$, $(A \times B) \times C$ and $A \times (B \times C)$. Discuss whether the two ordered pairs $((a_1,\ b_1),\ c_1)$ and $(a_1,\ (b_1,\ c_1))$ are equal. Discuss whether the two sets $(A \times B) \times C$ and $A \times (B \times C)$ are equal, or whether there is any sense in which they behave like one another.

3.3   RELATIONS

The relationships discussed in §3.1 and the elementary numerical functions such as the cosine function lead to sets of ordered pairs which fall broadly into three main types—equivalence relations, partial order relations and functions. These will be discussed in the next three sections. They have many aspects in common, and in this section a conceptual model of these common aspects is introduced.

In the model a *relation* is defined to be a set of ordered pairs. Readers who enjoy studying general concepts before particular examples will wish to look at the representation, inversion and composition of relations before moving on to the subsequent sections. Those who prefer to study the particular before the general will find it easier to skip this section and refer back to it when necessary.

Being a set of ordered pairs, a relation may be specified by the rule method or by the list method. The following relations will be used as examples throughout this section.

$$S_1 = \{(x, y) \mid x \in R \wedge y \in R \wedge y = x^2\},$$
$$S_2 = \{(x, y) \mid x \in R \wedge y \in R \wedge x < y\},$$
$$S_3 = \{(A, B) \mid B \subseteq U \wedge A \subseteq B\},$$
$$S_4 = \{(0, 1), (3, 4)\ (3, 6), (1, 0)\},$$
$$S_5 =. \{(a, r), (b, r)\ (c, s), (c, t)\},$$
$$S_6 = \{(1, 2), (2, 3), (3, 4), (4, 5), \ldots\}.$$

Given a relation S, the set of first components of the elements of S is called the *domain* of S, and is denoted by $\mathscr{D}(S)$. In the rule method of describing sets,

$$\mathscr{D}(S) = \{x \mid \exists\ y, (x, y) \in S\}.$$

Similarly, the set of second components of elements in S is called the *range* of S. The range of S is denoted by $\mathscr{R}(S)$, and we have

$$\mathscr{R}(S) = \{y \mid \exists\ x, (x, y) \in S\}.$$

Thus

$$\mathscr{D}(S_1) = \mathscr{D}(S_2) = \mathscr{R}(S_2) = R,$$

while

$$\mathscr{R}(S_1) = \{y \mid y \in R \wedge y \geqslant 0\}.$$

Also

$$\mathscr{D}(S_3) = \mathscr{R}(S_3) = \mathscr{P}(U), \qquad \mathscr{D}(S_4) = \{0, 1, 3\},$$

while

$$\mathscr{R}(S_4) = \{0, 1, 4, 6\}.$$

With these notations for domain and range, it follows that $S \subseteq \mathscr{D}(S) \times \mathscr{R}(S)$. More generally, if the elements of $\mathscr{D}(S)$ are specified as being members of some particular set A, and if the elements of $\mathscr{R}(S)$ are specified as being members of a set B, then $S \subseteq A \times B$, and we speak of S as being a relation from A to B. A relation from

A to A is said to be a relation *in* A. A relation S from A to A such that $\mathcal{D}(S) = A$ is said to be a relation *on* A.

The definition and terminology of relations introduced above will be found to be particularly valuable in the next sections. However, it does leave us with some problems of nomenclature in circumstances where the word relation has previously been used in a different sense. Thus in arithmetic we speak of $<$ as a relationship between numbers, whereas in our present discussion we speak of the set S_2 as the relation and call $x < y$ an open statement which defines the relation. This is, of course, a case of the chicken and the egg: with a given domain the relation and the open statement specify each other. Again, the relation S_3 is a relation on $\mathcal{P}(U)$ for some universal set U. It is natural to try to find an adjective to describe the relation. It is called the inclusion relation, and the name is sometimes applied ambiguously to the open statement $A \subseteq B$ which specifies the relation S_3. When relationships of kinship between people are being considered, it is difficult to avoid applying the word 'relation' to people rather than to the sets of pairs of people involved. It should be noted that words describing kinship are used ambiguously in everyday language, and before the conceptual model of sets of ordered pairs can be applied to problems involving kinship the ambiguities must be resolved. For example, it must be decided whether 'brother' is to include half-brother, whether 'father' is to include step-father, precisely what relatives are to be covered by the term 'cousin'.

A relation from a set A to a set B can be usefully represented in a diagram if A and B are sets of numbers, or if they have only a few elements. For example, the relation $S_1 \subset R^2$ can be represented by

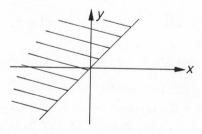

Figure 3.1. Graph of S_2

the familiar graph specified by the equation $y = x^2$. Similarly, the relation S_2 is represented by the shaded region in Fig. 3.1. The relations S_4 and S_6 can also be represented by graphs, though in these cases isolated points have to be indicated as in Fig. 3.2.

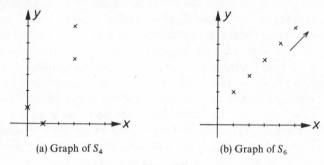

(a) Graph of S_4 (b) Graph of S_6

Figure 3.2

For the relation S_5 the domain is the set $\{a, b, c\}$ and the range is the set $\{r, s, t\}$. A diagram in the form of a lattice of points can be constructed by representing $\mathscr{D}(S_5)$ by points on a horizontal line and $\mathscr{R}(S_5)$ by points on a vertical line. The points which represent ordered pairs belonging to the relation can then be picked out by a cross or a circle to form a graph of the relation. On the other hand if $\mathscr{D}(S_5)$ and $\mathscr{R}(S_5)$ are represented by points on two lines parallel to each other, then the relation S_5 can be represented by the diagram of arrows of Fig. 3.3 (b). Such a representation of a relation will be called an arrow diagram.

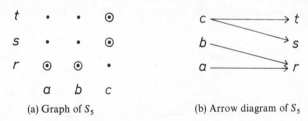

(a) Graph of S_5 (b) Arrow diagram of S_5

Figure 3.3

The definition of a relation embraces such widely differing ideas as '$y = x^2$' and 'he is her brother'. Each of these statements can be turned round, the first to '$x = \pm\sqrt{y}$' and the second to 'she is his sister'. Similarly, to an abstract relation S there corresponds a relation which consists of the reverses of the ordered pairs which belong to S. This relation is called the *inverse relation* to S, and is denoted by S^{-1}. It is specified using the rule method by

$$S^{-1} = \{(x, y) \mid (y, x) \in S\}.$$

Note that the domain of S is the range of S^{-1}, and the range of S is the domain of S^{-1}. Note also that if $S \subseteq R^2$ then the graph of S^{-1} is obtained by reflecting the graph of S in the line $y = x$; and an arrow diagram of an inverse relation T^{-1} is obtained by reversing the arrows in an arrow diagram of T.

Consider now the two statements '$y = \cos x^2$' and 'a is b's brother's son'. In the first case there must be a number z such that $z = x^2$ and $y = \cos z$. In the second statement there are three people involved, and the statement is an abbreviation for the statement 'a is c's son and c is b's brother'. Abstract relationships are similarly combined in the following way. If S and T are two relations then a relation $T \circ S$ is defined by

$$T \circ S = \{(x, y) \mid \exists z, (x, z) \in S \wedge (z, y) \in T\}.$$

Note that if $(x, z) \in S$ and $(z, y) \in T$ then $z \in \mathcal{R}(S) \cap \mathcal{D}(T)$, and $(x, y) \in \mathcal{D}(S) \times \mathcal{R}(T)$. Thus $T \circ S \subseteq \mathcal{D}(S) \times \mathcal{R}(T)$. If $\mathcal{R}(S) \cap \mathcal{D}(T) = \varnothing$ then $T \circ S = \varnothing$. The relation $T \circ S$ is called the *composition* of S and T.

As an example of a composed relation, let

$$S = \{(a, h), (b, h), (c, k)\}, \qquad T = \{(h, r), (k, s), (k, t)\},$$

so that $\mathcal{D}(S) = \{a, b, c\}$, $\mathcal{R}(S) = \mathcal{D}(T) = \{h, k\}$ and $\mathcal{R}(T) = \{r, s, t\}$. Then $(x, y) \in T \circ S$ is an open statement whose truth value relative to the truth of the statements which define S and T can be checked. Is $(a, r) \in T \circ S$ a true statement? To answer this we must check all the elements of $\mathcal{R}(S)$ to see whether '$(a, h) \in S \wedge (h, r) \in T$' or '$(a, k) \in S \wedge (k, r) \in T$' is true. The first of these is true, so that $(a, r) \in T \circ S$ is a true statement. Now $(a, s) \in T \circ S$ is true if '$(a, h) \in S \wedge (h, s) \in T$' or '$(a, k) \in S \wedge (k, s) \in T$' is true. Neither of the statements is true, and so $(a, s) \in T \circ S$ is not true. The other elements of $\mathcal{D}(S) \times \mathcal{R}(T)$ can be checked in a similar way to see whether or not they belong to the relation $T \circ S$, and thus it can be verified that

$$T \circ S = \{(a,r), (b,r), (c,s), (c,t)\}.$$

This can be illustrated most effectively by arrow diagrams. The relation $T \circ S$ is precisely the relation S_5 which may be represented by composing the arrows in the diagrams of S and T, as can be seen from Fig. 3.3 (b) and Fig. 3.4. The diagram would indicate that

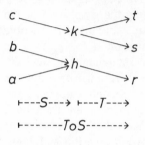

Figure 3.4

perhaps we should put the S before the T rather than use the notation $T \circ S$. However, the chosen notation conforms with the classical notations for functions such as $\log(\cos x)$ which are evaluated from right to left.

We have defined a relation as a set of ordered pairs, and so the empty set is a relation, according to this definition. We have shown in Chapter 2 that the empty set is a subset of each set under discussion, and therefore, given any two sets A and B, the empty set $\varnothing$ is a relation from A to B. So is the whole set $A \times B$, though this relation like the empty set relation will be useful in the general theory of relations rather than in any particular application.

Among the relations in a set A, one which is of particular importance is the relation

$$\{(x,y) \,|\, x \in A \text{ and } x = y\},$$

which is called the *identity relation* on A and is denoted by I_A. Thus for each element $x \in A$, we have $(x,x) \in I_A$; and if x and y are distinct then $(x,y) \notin I_A$. If S is a relation from A to B then $I_B \circ S = S$, and $S \circ I_A = S$.

Exercises 3.3

1. For the relations T_1 to T_8 of exercise 1 of §3.2 write down the inverse relations and sketch their graphs.

2. Let $S = \{(a,f),(a,g),(b,h),(c,f),(e,g),(e,h)\}$,
   $\quad T = \{(f,p),(g,q),(g,r),(g,t),(h,p),(h,s),(h,t)\}$.
Write down all the elements of the composite relation $T{\circ}S$. Draw diagrams of the relations S, T and $T{\circ}S$.

3. Let S and T be relations defined in R as follows:
$$S = \{(x,y)\,|\,y = (x+1)^3\}, \qquad T = \{(x,y)\,|\,y = 2x-2\}.$$
Find the relations $T{\circ}S$ and $S{\circ}T$. Draw diagrams of all four relations.

4. Let T_1 to T_8 denote the relations of exercise 1 of §3.2. Write down the following relations in rule form: $T_1{\circ}T_3$, $T_3{\circ}T_1$, $T_2{\circ}T_3$, $T_3{\circ}T_2$, $T_2{\circ}T_4$, $T_4{\circ}T_2$, $T_4{\circ}T_8$, $T_8{\circ}T_4$.

5. Let S and T be relations defined in R as follows:
$$S = \{(x,y)\,|\,y = Ax+B\}, \qquad T = \{(x,y)\,|\,y = Cx+D\}.$$
Find $T{\circ}S$ and $S{\circ}T$, and discuss circumstances under which $T{\circ}S = S{\circ}T$.

3.4 EQUIVALENCE RELATIONS

Of the special kinds of relations which will be studied in the rest of this chapter, the first which we shall consider is that associated with a subdivision of a set into disjoint subsets. Such subdivisions occur frequently in mathematics and also in many other situations involving classification. For example, the set of towns in England is subdivided by counties. Each town is in precisely one county. Any ambiguity about which places are to be regarded as towns can be resolved by reference to a chosen gazetteer, and a map of England which shows towns and county boundaries can be used as an illustration of this subdivision of the set of towns.

The formal set theoretic notion which corresponds to this idea of subdivision is that of a partition. A *partition* of a set A is a collection $\mathscr{C}$ of subsets of A which satisfies the following three conditions:

P_1 : if $X \in \mathscr{C}$, $Y \in \mathscr{C}$ and $X \neq Y$, then $X \cap Y = \varnothing$;

P_2: $\forall x \in A, \exists X \in \mathscr{C}, x \in X$;

P_3: $\varnothing \notin \mathscr{C}$.

By using the law of contrapositive the first condition can be written in the alternative form: if $X \in \mathscr{C}$, $Y \in \mathscr{C}$ and $X \cap Y \neq \varnothing$, then $X = Y$.

The expression 'the subsets are *pairwise disjoint*' is sometimes used to describe condition P_1, while 'the subsets form a *cover*' is sometimes used to describe condition P_2. So a partition is a cover by pairwise disjoint non-empty subsets. The three conditions ensure that each element x of A belongs to precisely one of the subsets in the partition. A notation such as A_x is frequently used for the subset to which an element x belongs. This notation corresponds to the statement 'the county in which Southampton is situated'. We know that this county has the special name Hampshire, yet we may not have wished to use it. On the other hand, many readers will not know the name of the county in which Dent is situated, and so will not be able to use it. In some mathematical situations in which partitions are used, it is not necessary to give special names to the subsets involved; they can be studied effectively by the use of notations like A_x. On the other hand, the subsets in a partition may be given special names such as K, L, M, or $A_1, A_2, \ldots, A_n$, or more generally A_i where the domain of the suffix i is some indexing set I. In the latter notation the property P_2 can be expressed as

$$A = \bigcup_{i \in I} A_i.$$

We shall now show that associated with a partition of a set A there is a relation which has three properties, and that every relation with these three properties gives rise to a partition. Readers may find it helpful while working through the remainder of this section to keep in mind the relation between pairs of towns of being in the same county.

Given a partition $\mathscr{C}$ of a set A, define a relation

$$S = \{(x, y) \mid \exists X \in \mathscr{C}, x \in X \text{ and } y \in X\}.$$

We shall prove that S has the following three properties which will in future be referred to by the names given to them here.

Reflexive: $\forall x \in \mathscr{D}(S), (x, x) \in S$.

Symmetric: if $(x, y) \in S$, then $(y, x) \in S$.

Transitive: if $(x, y) \in S$ and $(y, z) \in S$, then $(x, z) \in S$.

The proofs consist of demonstrating that if the statements P_1, P_2 and P_3 and the statement which defines the relation are given the truth value T then the three statements in question have the truth value T relative to the basic statements of set theory.

The proofs are presented in an abbreviated formal style which reflects the transitivity of equivalence and of implication (tautology ix). It is left as an exercise to the reader to write out the proofs in discursive style.

Proof of reflexivity

$$\forall x \in \mathcal{D}(S), (x, x) \in S \Leftrightarrow \forall x \in \mathcal{D}(S), \exists X \in \mathscr{C}, x \in X \wedge x \in X$$
$$\text{(definition of } S)$$

$$\Leftrightarrow \forall x \in \mathcal{D}(S), \exists X \in \mathscr{C}, x \in X$$
$$\text{(idempotence, tautology iv)}$$

The last statement is P_2 with $\mathcal{D}(S)$ for A; hence the relation S is reflexive, and $\mathcal{D}(S) = A$.

Proof of symmetry

$$(x, y) \in S \Leftrightarrow \exists X \in \mathscr{C}, x \in X \wedge y \in X$$
$$\text{(definition of } S)$$

$$\Leftrightarrow \exists X \in \mathscr{C}, y \in X \wedge x \in X$$
$$\text{(commutativity, tautology i)}$$

$$\Leftrightarrow (y, x) \in S \qquad \text{(definition of } S)$$

Thus $(x, y) \in S \Rightarrow (y, x) \in S$, i.e. the relation is symmetric.

Proof of transitivity

$$((x, y) \in S \wedge (y, z) \in S)$$
$$\Leftrightarrow ((\exists X \in \mathscr{C}, x \in X \wedge y \in X) \wedge (\exists Y \in \mathscr{C}, y \in Y \wedge z \in Y))$$
$$\text{(definition of } S)$$

Now

$$(y \in X) \wedge (y \in Y) \Rightarrow X \cap Y \neq \varnothing,$$
$$(X \in \mathscr{C} \wedge Y \in \mathscr{C} \wedge X \cap Y \neq \varnothing) \Rightarrow X = Y$$
$$\text{(contrapositive form of } P_1)$$

Thus

$$((x, y) \in S \wedge (y, z) \in S) \Rightarrow \exists X \in \mathscr{C}, x \in X \wedge y \in X \wedge z \in X$$

$$\Rightarrow \exists X \in \mathscr{C}, x \in X \wedge z \in X$$

(simplification, tautology vii)

$$\Rightarrow (x, z) \in S \qquad \text{(definition of } S)$$

i.e. the relation is transitive.

Each of the first two proofs consists of a short sequence of remarks, each one justified by a definition or a tautology. In the third proof the flow of the main argument is interrupted by the intrusion of the subsidiary argument which proves that $X = Y$. This proof illustrates a point about the use of variables. In the first sentence of the proof, having used the symbol X for the subset which contains x and y, we cannot immediately use the same symbol for the subset which contains y and z. Such a subset is known to exist, and the symbol Y is chosen for it. It is then proved that X and Y are different symbols for the same set.

Now we shall show that a relation S which is reflexive, symmetric and transitive gives rise to a partition of $\mathscr{D}(S) = A$. Given such a relation, for each element x of A we consider the set $A_x = \{y \mid (x, y) \in S\}$. We shall show that the collection $\{A_x \mid x \in A\}$ of subsets of A satisfies properties P_1, P_2 and P_3 and so is a partition of A. This time the proofs are presented in discursive style and it is left to each reader to clarify the logic of the proofs to the extent that he finds necessary by setting out a more formal proof.

Proof of P_2 and P_3 Since S is reflexive, given an arbitrary element x of A, we have $(x, x) \in S$. It follows from definition of A_x that $x \in A_x$. Thus $\{A_x \mid x \in A\}$ has properties P_2 and P_3.

While the proof of these properties is trivial, the proof of P_1 given below uses proof by parts applied to the contrapositive of P_1. The assumption $A_x \cap A_y \neq \varnothing$ is shown below to imply that $A_x \subseteq A_y$ and that $A_y \subseteq A_x$; so by the tautology of proof by parts the assumption $A_x \cap A_y \neq \varnothing$ implies that $A_x = A_y$. Since the proof that $A_x \cap A_y \neq \varnothing$ implies $A_y \subseteq A_x$ is a repetition of the proof that $A_x \cap A_y \neq \varnothing$ implies $A_x \subseteq A_y$ with no more that some interchanges of the symbols x and y, the details of this step are omitted.

Proof of P_1 Suppose that $A_x \cap A_y \neq \varnothing$. Then there exists $z \in A$

such that $z \in A_x$ and $z \in A_y$. Let w be an arbitrary element in A_x. Then $(x, z) \in S$, $(y, z) \in S$ and $(x, w) \in S$. Since S is symmetric, $(z, x) \in S$; and so, since S is transitive, $(z, w) \in S$ and $(y, w) \in S$; whence $w \in A_y$. Since w was arbitrary, $A_x \subseteq A_y$. Similarly $A_y \subseteq A_x$. It follows by proof by parts that if $A_x \cap A_y \neq \varnothing$ then $A_x = A_y$.

We have completed the demonstration that partitions and the special type of relation studied here give rise to each other. Thus relations which are reflexive, symmetric and transitive have the effect of grouping together elements which are in some sense equivalent to each other. They are commonly called *equivalence relations*. If S is an equivalence relation and $(x, y) \in S$ then x and y are said to be *S-equivalent*, or simply *equivalent* if only one equivalence relation is under discussion. An alternative notation for this is $x \, S \, y$. The partition of a set A associated with an equivalence relation S on A is called the *quotient set* of the relation and is denoted by A/S. The subset in the partition to which an element x of A belongs is called the *equivalence class* of x. It may be denoted by A_x or by $[x]$.

To check whether a relation S is an equivalence relation, it is necessary to check whether it is reflexive and symmetric and transitive. For the first, all the elements of S of the form (x, x), where $x \in \mathscr{D}(S) = A$, must be shown to be in S, i.e. it must be shown that $I_A \subseteq S$. For the second property all elements of S of the form (x, y) must be checked, while for the third all ordered pairs of elements of S of the form $((x, y), (y, z))$ must be checked. For example, let $A = \{a, b, c\}$ where a, b and c are distinct, and let

$$S_7 = \{(a, a), (b, b), (c, c), (a, b), (a, c), (c, b)\},$$
$$S_8 = \{(a, a), (b, b), (a, b), (a, c), (b, a), (c, a)\}.$$

Then S_7 is reflexive because $\mathscr{D}(S_7) = A$, $(a, a) \in S_7$, $(b, b) \in S_7$ and $(c, c) \in S_7$. However, S_8 is not reflexive because $c \in \mathscr{D}(S_8)$ while $(c, c) \notin S_8$. By checking that for each of the six elements of S_8, the pair in reverse order is also in S_8, we see that S_8 is symmetric. On the other hand, to demonstrate that S_7 is not symmetric we need display only one element of S_7 for which the pair in reverse order is not in S_7, and (a, c) is such an element. The full list of pairs of elements of S_7 of the form $((x, y), (y, z))$ is found by inspection to be

$$((a,a),(a,a)), \quad ((a,a),(a,b)), \quad ((a,a),(a,c)),$$
$$((b,b),(b,b)), \quad ((c,c),(c,c)), \quad ((c,c),(c,b)),$$
$$((a,b),(b,b)), \quad ((a,c),(c,c)), \quad ((a,c),(c,b)),$$
$$((c,b),(b,b)).$$

The ten elements (x,z) corresponding to these ten ordered pairs of elements are

$$(a,a), \quad (a,b), \quad (a,c), \quad (b,b), \quad (c,c),$$
$$(c,b), \quad (a,b), \quad (a,c), \quad (a,b), \quad (c,b).$$

Since each of these is an element of S_7, the relation S_7 is transitive. On the other hand, we demonstrate that S_8 is not transitive by displaying a pair of elements of S_8 of the form $((x,y),(y,z))$ for which the corresponding element (x,z) does not belong to S_8. Now $(b,a) \in S_8$ and $(a,c) \in S_8$ while $(b,c) \notin S_8$. Hence S_8 is not transitive. It follows that neither S_7 nor S_8 is an equivalence relation. On the other hand the relation $S_9 = I_A \cup \{(a,c),(c,a)\}$ is an equivalence relation with equivalence classes $\{a,c\}$ and $\{b\}$. The quotient set A/S_9 is the set $\{\{a,c\},\{b\}\}$.

When the set $\mathscr{D}(S)$ and the relation S have large numbers of elements the above method of showing that S is reflexive, symmetric or transitive cannot be used, although it is still true that single counterexamples serve to show that S does not have the properties. In this case the relation will generally be defined in terms of some open statement which can be manipulated to show whether or not S is an equivalence relation. The following example is of this type. It is generalized in exercise 1(a) below.

Let the relation S be defined on the set Z of integers by the statement

$$S = \{(x,y) \mid x-y \text{ is divisible by 3}\}.$$

To show that S is an equivalence relation we must prove that it is reflexive, symmetric and transitive. Note that the statement '$x-y$ is divisible by 3' can be written using a quantifier as '$\exists n \in Z, x-y = 3n$' and we shall use this form of the statement in the proofs of the three properties.

Proof of reflexivity

$$\forall x \in Z, \quad x - x = 0 = 0.3.$$

Thus $\forall x \in Z$, $x - x$ is divisible by 3, i.e.

$$\forall x \in Z, \quad (x, x) \in S.$$

Proof of symmetry Suppose that $(x, y) \in S$. Then $x - y$ is divisible by 3, i.e. $\exists n \in Z$, $x - y = 3n$. Hence $y - x = 3(-n)$, i.e. $y - x$ is divisible by 3, and therefore $(y, x) \in S$.

Proof of transitivity Suppose $(x, y) \in S$ and $(y, z) \in S$. Then $x - y$ and $y - z$ are both divisible by 3, i.e. $\exists m, n \in Z$, $x - y = 3m$ and $y - z = 3n$. Adding these two equations gives $x - z = 3(m + n)$, i.e. $x - z$ is divisible by 3 and so $(x, z) \in S$.

To discuss the associated partition into equivalence classes, we notice that the integer 0 is related by S to 3, -3, 6, -6, etc., the integer 1 is related to 4, -2, 7, -5, etc., and the integer 2 is related to 5, -1, 8, -4, etc. Together these sets of integers comprise the whole of Z, and so we have just three equivalence classes,

$$A_0 = \{0, 3, -3, 6, -6, 9, -9, \ldots\},$$
$$A_1 = \{1, 4, -2, 7, -5, 10, -8, \ldots\},$$
$$A_2 = \{2, 5, -1, 8, -4, 11, -7, \ldots\}.$$

The quotient set Z/S is therefore the set $\{A_0, A_1, A_2\}$.

These examples illustrate the fact that the elements of the quotient set of an equivalence relation on a set A are not simply elements of A but *sets* of elements of A. The basic idea is that we are interested in only one aspect of the elements of a set, and so we collect together all those elements which look the same when considered from this aspect. Thus certain days which have one aspect in common with each other are all called 'Monday'. The word 'Monday' is the collective noun for such typical Mondays as 'last Monday' and 'the second Monday in January' and 'Monday, 30 March 1970'. There is an underlying equivalence relation, two days being equivalent if one follows the other by a multiple of seven days. This equivalence relation has seven equivalence classes, which are given the names of the days of the week. The seven equivalence classes are the elements of the quotient set of the relation.

The mathematicians who first defined the concept of an equivalence relation did so in an attempt to produce a mathematical model of situations involving classification. They thought that the conditions of symmetry and transitivity were sufficient. Only later was it seen that reflexivity was also needed to give the connection we have described between equivalence relations and partitions. The reasons why they did not consider the reflexive property become clear as soon as one looks at an example. We considered at the beginning of this section the subdivision of towns by counties. It would not normally occur to one to ask whether a town is in the same county as itself. Similarly, in dealing with relations of kinship, one would not normally ask such questions as 'Is John the brother of himself?' or 'Has John the same father as himself?'. It follows in consequence that when considering the question of transitivity one would study only triples of distinct elements. Implications of the type 'if $(x, y) \in S$ and $(y, x) \in S$ then $(x, x) \in S$' would not be regarded as significant in everyday conversations about kinship. However, the following example will show the need for the reflexive property and for the full strength of the transitive property if the equivalence classes are to form a partition.

Let $A = \{a, b, c, d, e, f\}$, and let

$$S = \{(a, b), (b, a), (b, c), (c, b), (a, c), (c, a), (d, d), (d, e), (e, d)\} .$$

It is easy to see that S is symmetric. Since $(a, b) \in S$ and $(b, a) \in S$ while $(a, a) \notin S$, it follows that S is not transitive. Nevertheless, it can be checked that S satisfies a weak form of the transitive property, namely, 'if $(x, y) \in S$ and $(y, z) \in S$ and $x \neq z$, then $(x, z) \in S$'. Now when we form the sets $A_x = \{y \mid (x, y) \in S\}$ we obtain $A_a = \{b, c\}$, $A_b = \{c, a\}$, $A_c = \{b, a\}$, $A_d = \{d, e\}$, $A_e = \{d\}$, $A_f = \varnothing$. Thus the attempt to classify the elements of A by means of this relation does not lead to a partition. Indeed, it is easy to see that P_1, P_2 and P_3 are each violated by the collection of sets $\{A_x \mid x \in A\}$.

In contrast to the rather awkward situation exemplified here, equivalence relations and partitions are very easy to work with, and so give rise to satisfactory models for many mathematical and everyday situations.

Exercises 3.4

1. Show that the relation S defined on the set $A = \mathscr{D}(S)$ is an equivalence relation in each of the cases described below. In each case describe the quotient set A/S.

(a) $A = Z$, the set of integers, and $(a, b) \in S$ if and only if a and b have the same remainder on division by m, where m is a fixed positive integer, or equivalently, if and only if $a - b$ is a multiple of m. There is a special terminology and notation for this equivalence relation. If $(a, b) \in S$, then a is said to be congruent to b modulo m, and the notation $a \equiv b \pmod{m}$ is used.

(b) A is the set of symbols of the form a/b where $a \in Z$, $b \in Z$ and $b \neq 0$, and $(a/b, c/d) \in S$ if and only if $ad = bc$.

(c) A is the set N^2 and $((a, b), (c, d)) \in S$ if and only if $a + d = b + c$.

(d) $A = C$, the set of complex numbers, and $(z, w) \in S$ if and only if $|z| = |w|$.

(e) $A = P$, the set of polynomials with real coefficients, and $(p(x), q(x)) \in S$ if and only if

$$\frac{d}{dx} p(x) = \frac{d}{dx} q(x).$$

2. (a) Show that if S and T are equivalence relations on a set $A = \mathscr{D}(S) = \mathscr{D}(T)$, then $T \circ S$ is an equivalence relation if and only if $T \circ S = S \circ T$.

(b) Investigate $T \circ S$ when A is the set of integers, S is the relation of congruence modulo 6, and T the relation of congruence modulo 4. By means of further numerical examples, or by any other method, try to conjecture and then prove a result concerning an arbitrary pair of moduli.

3. Show that if S is an equivalence relation then S^{-1} is also an equivalence relation.

3.5 PARTIAL ORDER RELATIONS

The relation $T_1 = \{(x, y) \mid x \leqslant y\}$ on the set R of real numbers, and the relation $T_2 = \{(X, Y) \mid X \subseteq Y\}$ on the power set $\mathscr{P}(U)$ of a non-empty set U are reflexive and transitive. Since $(1, 2) \in T_1$ while $(2, 1) \notin T_1$, the relation T_1 is not symmetric. Also the relation T_2 is not symmetric, for $(\varnothing, U) \in T_2$ whilst $(U, \varnothing) \notin T_2$. Indeed, if $x \leqslant y$ and $y \leqslant x$ then $x = y$, while if $X \subseteq Y$ and $Y \subseteq X$ then $X = Y$. Thus both T_1 and T_2 satisfy the following property which is stated here for a general relation S.

Anti-symmetry: if $(x, y) \in S$ and $(y, x) \in S$ then $x = y$.

Note that symmetry and anti-symmetry are not negations of one another. If the elements a, b, and c are distinct, then the relation $\{(a,b),(a,c),(c,a)\}$ is neither symmetric nor anti-symmetric, while the relation $\{(a,a),(b,b)\}$ is both symmetric and anti-symmetric.

A relation which is reflexive, anti-symmetric and transitive is said to be a *partial order relation*.

The relation T_1 is distinguished from the relation T_2 in that given any pair of real numbers x,y either $x \leqslant y$ or $y \leqslant x$, while so long as U has at least two elements there exist subsets X, Y such that neither is a subset of the other. A partial order relation which satisfies the extra condition

$$\forall x, y \in \mathscr{D}(S), (x,y) \in S \lor (y,x) \in S$$

is said to be a *full ordering* or a *total ordering* or a *linear ordering*.

Conditions on a relation which generalize the ideas of strict inequality for real numbers and of strict inclusion for subsets can be obtained in a similar way. However, there is no generally accepted language for the conditions which arise in that case. It is therefore convenient to follow mathematical tradition and use partial order relations as described above.

For relations which have few elements the process of checking whether they are partial order relations is similar to the process of checking whether they are equivalence relations. Consider the examples S_7, S_8 and S_9 of the previous section. Since the only pairs of elements of the form (x,y), (y,x) in S_7 are (a,a), (a,a); (b,b), (b,b); (c,c), (c,c); the relation S_7 is anti-symmetric. On the other hand both S_8 and S_9 contain the pair (a,c), (c,a), so neither of these relations is anti-symmetric. Thus S_8 and S_9 are not partial order relations. Since S_7 is reflexive, anti-symmetric and transitive, it is a partial order relation. Indeed it is a total ordering of the three elements in the order a then c then b.

The method of showing that a relation defined by an open statement is a partial order relation corresponds to that illustrated at the end of the last section for equivalence relations. As an example we consider the relation S defined on the set A of polynomials as follows.

$$S = \{(p_1, p_2) \,|\, \exists q \in A, \forall x \in R, q(x) \geqslant 1 \land p_1(x) = q(x)p_2(x)\}$$

To show that S is a partial order relation we have to show that it is reflexive, anti-symmetric and transitive.

Proof of reflexivity   Let i denote the polynomial such that $\forall x \in R$, $i(x) = 1$. Then

$$\forall p \in A, i(x) \geqslant 1 \wedge p(x) = i(x)p(x), \qquad \text{i.e. } \forall p \in A, (p, p) \in S.$$

Proof of anti-symmetry   Suppose $(p_1, p_2) \in S$ and $(p_2, p_1) \in S$. Then

$$\exists q_1 \in A, \quad \forall x \in R, \quad q_1(x) \geqslant 1 \wedge p_1(x) = q_1(x)p_2(x)$$

$$\exists q_2 \in A, \quad \forall x \in R, \quad q_2(x) \geqslant 1 \wedge p_2(x) = q_2(x)p_1(x).$$

Thus

$$\forall x \in R, \quad p_1(x) = q_1(x)q_2(x)p_1(x),$$

i.e.

$$\forall x \in R, \quad p_1(x) = p_2(x) = 0 \vee q_1(x)q_2(x) = 1.$$

But

$$\forall x \in R, \quad q_1(x) \geqslant 1 \wedge q_2(x) \geqslant 1$$

and so

$$\forall x \in R, \quad p_1(x) = p_2(x) = 0 \vee q_1(x) = q_2(x) = 1.$$

Thus

$$\forall x \in R, \quad p_1(x) = p_2(x), \qquad \text{i.e. } p_1 = p_2.$$

Proof of transitivity   Suppose $(p_1, p_2) \in S$ and $(p_2, p_3) \in S$. Then

$$\exists q_1 \in A, \forall x \in R, q_1(x) \geqslant 1 \wedge p_1(x) = q_1(x)p_2(x),$$

$$\exists q_2 \in A, \forall x \in R, q_2(x) \geqslant 1 \wedge p_2(x) = q_2(x)p_3(x).$$

Thus

$$\forall x \in R, p_1(x) = q_1(x)q_2(x)p_3(x).$$

But $q_1 q_2 \in A$ and $\forall x \in R, q_1(x)q_2(x) \geqslant 1$, and so

$$(p_1, p_3) \in S.$$

Thus S is a partial order relation.

A visual example of a partial order relation is afforded by the branches of a tree. If we think of the branching points as elements of a set then a relation on that set which consists of ordered pairs of points (a, b) such that b is on one of the branches which comes from a is a partial order relation relation. Every partial relation on a set with only a small number of elements can be convenient represented by a branching diagram. We represent the elements of the set A as points, and the diagram is constructed by joining together 'adjacent' points by a directed line segment. The complete specification of the partial order relation S can then be deduced from the diagram by using reflexivity and transitivity. When we say that a and b are 'adjacent' points in the partially ordered set we mean that $a \neq b$, $(a, b) \in S$ and there is no element $c \in A$ with the property that $(a, c) \in S$ and $(c, b) \in S$, i.e. there is no element c 'between' a and b. Thus Fig. 3.5(a) is a branching diagram for the partial order relation

$$I_A \cup \{(a, b), (a, c), (a, d), (a, e), (c, d), (c, e)\}$$

on the set $A = \{a, b, c, d, e\}$. For some purposes this is more convenient than the graph or the arrow diagram of the relation which are shown in Fig. 3.5(b), (c). Fig. 3.6 represents a partial order relation with 39 elements on a set with 12 elements. The graph and the arrow diagram would both be cumbersome, and it would not be at all clear that they represented a partial order relation

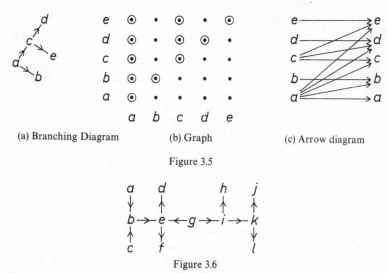

(a) Branching Diagram   (b) Graph   (c) Arrow diagram

Figure 3.5

Figure 3.6

Exercises 3.5

1. Write down all the elements of the partial order relation whose branching diagram is that of Fig. 3.6.

2. Show that the relation S defined on the set $A = \mathcal{D}(S)$ is a partial order relation in each of the cases described below. Are any of the relations total order relations?

(a) $A = \mathcal{P}(U)$ for some set U. $(B, C) \in S$ if and only if $C \subseteq B$.

(b) $A = C$, the set of complex numbers. $(z_2, z_2) \in S$ if and only if $|z_1| \leqslant |z_2|$ and $\arg z_1 = \arg z_2$.

(c) $A = N$, the set of natural numbers. $(m, n) \in S$ if and only if $m \leqslant n$.

(d) $A = R \times R$, where R is the set of real numbers. $((x, y), (x', y')) \in S$ if and only if either $x < x'$, or $x = x'$ and $y \leqslant y'$.

(e) $A = \{1, 2, 3, 5, 6, 10, 15, 30\}$. $(m, n) \in S$ if and only if n is divisible by m.

(f) $A = \{1, 2, 3, 5, 10, 15\}$. $(m, n) \in S$ if and only if n is divisible by m.

(g) $A = \{1, 2, 4, 8, 16, 32\}$. $(m, n) \in S$ if and only if n is divisible by m.

3. S is a partial order relation on the set $A = \mathcal{D}(S)$. Is S^{-1} a partial order relation on A?

4. Suppose that we have a set A with a partial order relation S. Let $a \in A$ and $b \in A$. An element $c \in A$ is said to be a *least upper bound* (l.u.b.) for the set $\{a, b\}$ if and only if

(i) $(a, c) \in S$ and $(b, c) \in S$,

(ii) $\forall d, (a, d) \in S$ and $(b, d) \in S$ implies $(c, d) \in S$.

An element g is said to be a *greatest lower bound* (g.l.b.) for the set $\{a, b\}$ if and only if

(i) $(g, a) \in S$ and $(g, b) \in S$,

(ii) $\forall h, (h, a) \in S$ and $(h, b) \in S$ implies $(h, g) \in S$.

Such elements c and g may or may not exist. Need they be unique if they do exist? A partially ordered set A such that every subset of the form $\{a, b\}$ possesses a g.l.b. and a l.u.b. is called a *lattice*.

Investigate this situation with respect to the partial order relations in exercise 2. If they are all lattices find a partially ordered set which is not a lattice. What can you say with respect to lattice properties about partially ordered sets which are totally ordered?

5. Let S be the equivalence relation on C defined in exercise 1(d) of §3.4. Let T be the partial order relation on C defined in exercise 2(b) above. If $[z]$ denotes the equivalence class containing z, we define a relation T^* on C/S by the rule

$$T^* = \{([z], [w]) \mid \exists z_0 \in [z], w_0 \in [w] \text{ such that } (z_0, w_0) \in T\}.$$

Show that T^* is a partial order relation on C/S. (T^* is said to be induced by T on C/S.)

3.6 FUNCTIONS

In this section we shall be concerned with a definition of functions which will be a foundation for the study of functions used in calculus such as the trigonometric and logarithmic functions, and for the study of functions used in algebra and all other branches of mathematics. The polynomial functions, and sin, tan, exp, log, etc., can all be looked upon as rules which associate with each value of an independent variable x precisely one value of a dependent variable y. In plotting the graph of such a function we make use of the ordered pairs (x, y). If we think of such functions as sets of ordered pairs then they can be regarded as relations.

The particular property of these relations with which we shall be concerned is that to each first coordinate x there corresponds precisely one second coordinate y. An alternative way of saying this is that if y and z are both second coordinates for the same first coordinate x, then y and z must be different symbols for the same element. We shall restrict ourselves further to relations in which the domain of the relation from A to B is the set A itself. This is a somewhat arbitrary restriction which is chosen in preference to the alternative of describing the set A and the domain of the function separately for each example. Thus we define a *function* from a set A to a set B as a relation S from A to B such that

F_1: $\mathscr{D}(S) = A$;

F_2: if $(x, y) \in S$ and $(x, z) \in S$, then $y = z$.

The condition F_2 is the crucial property which distinguishes functions from other kinds of relations. We insist on F_1 for technical reasons; it is sometimes useful for distinguishing between different functions. We do not insist on the analogous property that the range of S should be B, since if the function is defined by a complicated equation it may be difficult to discover precisely what the range is. For example, consider the equation

$$y = \max \{t \mid t^5 + xt^3 + t + 1 = 0\}.$$

For most particular real numbers x_0 substituted for the variable x, it will not be easy to solve the equation $t^5 + x_0 t^3 + t + 1 = 0$, and so find its largest root. Thus, given a set of real numbers as domain, it will be difficult to determine the range of the function defined in this way.

Some relations of kinship give rise to functions. For example, if A denotes a set of people and B denotes the set of their parents, then the relation $\{(x, y) \mid y \text{ is the father of } x\}$ from A to B is a function. As an example of a function which arises in calculus, let A be the set of polynomials $p(x)$, $q(x)$ etc. Then

$$\left\{ (p(x), q(x)) \mid q(x) = \frac{d}{dx} p(x) \right\}$$

is a function from A to itself.

We shall follow tradition in using letters f, g, h for functions, in writing $y = f(x)$ as an alternative for $(x, y) \in f$, and in calling $f(x)$ the image of x under f. We also write $f \colon A \rightsquigarrow B$ to denote that f is a function from A to B. Alternative words which are sometimes used for function are map, mapping and transformation. Functions are often defined by open statements. For example $f = \{(x, y) \mid x \in R \text{ and } y = x^2\}$ is a function from R to R. An alternative notation for this function is $f \colon R \rightsquigarrow R$, $f \colon x \rightsquigarrow x^2$. Note that by itself the open statement $y = x^2$ is not in our sense a function, nor does it define a unique function, but the open statement together with any subset of the set of real numbers does define a function. The function $f \colon R \rightsquigarrow R$, $f \colon x \rightsquigarrow x^2$ is not the same as the function $f \colon N \rightsquigarrow N$, $f \colon x \rightsquigarrow x^2$. Indeed, the open statement $y = x^2$ can be used to define functions on other domains such as a set of complex numbers, or a set of 2×2 matrices.

You will frequently find yourself having to discover the set theoretic functions underlying such statements as 'the function $y = \log x$' and 'the function $y^2 = x$'. A natural domain for a function defined by the open statement $y = \log x$ is the set of positive real numbers. With this domain the corresponding range would be the set of real numbers. Thus a function defined by $y = \log x$ is $f = \{(x, y) \mid x \in R, x > 0 \text{ and } y = \log x\}$ for which an alternative notation is

$$f \colon \{x \mid x \in R \text{ and } x > 0\} \rightsquigarrow R, f \colon x \rightsquigarrow \log x.$$

Different functions would be obtained by the choice of smaller domains.

The second open statement $y^2 = x$ gives more difficulty. Since $2^2 = 4$ and $(-2)^2 = 4$, we shall have to choose the set B as well as the set A carefully if the condition F_2 is to be satisfied by a relation defined by this open statement. We can take A to be the set of non-negative real numbers and B to be the same. Then

$$g = \{(x, y) \,|\, x \geqslant 0 \text{ and } y \geqslant 0 \text{ and } y^2 = x\},$$

$$g: \{x \,|\, x \in R \text{ and } x \geqslant 0\} \rightsquigarrow \{x \,|\, x \in R \text{ and } x \geqslant 0\}, \quad g: x \rightsquigarrow \sqrt{x}$$

are two alternative notations for a relation which satisfies properties F_1 and F_2 and so is a function. On the other hand by choosing the non-positive real numbers as the range we obtain the function

$$h = \{(x, y) \,|\, x \geqslant 0 \text{ and } y \leqslant 0 \text{ and } y^2 = x\},$$

$$h: \{x \,|\, x \in R \text{ and } x \geqslant 0\} \rightsquigarrow \{x \,|\, x \in R \text{ and } x \leqslant 0\}, \quad h: x \rightsquigarrow -\sqrt{x}.$$

At one time the union of g and h was referred to as the multivalued function $y^2 = x$ and it was said that g and h were single-valued functions and were the branches of the multivalued function. This terminology is obsolescent: multivalued functions are now usually regarded simply as relations, and all functions are single-valued.

The composition of two relations was defined in § 3.3. The definition applied to a function $f: A \rightsquigarrow B$ and a function $g: B \rightsquigarrow C$ is

$$g \circ f = \{(x, z) \,|\, \exists y, y = f(x) \text{ and } z = g(y)\}.$$

As was pointed out in §3.3, the reason for the order used in the notation $g \circ f$ is that it corresponds to the notation $z = g\,f(x))$. It is easy to convince oneself that $g \circ f$ is a function from A to C, i.e. that the domain of $g \circ f$ is A and that to each element of A there corresponds precisely one element of C. However, it is not easy to see how to handle the variables involved in proofs of these statements. In the proof given below x is an element of A, while y, y_1, y_2 are elements of B and z_1, z_2 are elements of C. In this notation the forms of condition F_2 for f, g and $g \circ f$ are respectively

(i) if $(x, y_1) \in f$ and $(x, y_2) \in f$, then $y_1 = y_2$;

(ii) if $(y, z_1) \in g$ and $(y, z_2) \in g$, then $z_1 = z_2$;

iii) if $(x, z_1) \in g \circ f$ and $(x, z_2) \in g \circ f$, then $z_1 = z_2$.

We have to prove that if the statement defining $g \circ f$ and statements (i) and (ii) have the truth value T then statement (iii) has the truth value T. The proof is as follows. Suppose that $(x, z_1) \in g \circ f$ and $(x, z_2) \in g \circ f$. Then by the definition of $g \circ f$, there exists $y_1 \in B$ such that $(x, y_1) \in f$ and $(y_1, z_1) \in g$, and there exists $y_2 \in B$ such that $(x, y_2) \in f$ and $(y_2, z_2) \in g$. Since $(x, y_1) \in f$ and $(x, y_2) \in f$ it follows that $y_1 = y_2$. Thus $(y_1, z_1) \in g$ and $(y_1, z_2) \in g$, whence $z_1 = z_2$. This completes the proof. The proof that the domain of $g \circ f$ is the set A is similar and is left as an exercise to the reader.

3.7 SPECIAL FUNCTIONS

A function f from a set A to a set B has been defined as a relation from A to B which satisfies the two conditions

$F_1 : \mathscr{D}(f) = A$;

$F_2 :$ if $(x, y) \in f$ and $(x, z) \in f$, then $y = z$.

We shall now classify functions further according to whether or not they satisfy the two additional conditions

$F_3 : \mathscr{R}(f) = B$;

$F_4 :$ if $(x, y) \in f$ and $(z, y) \in f$ then $x = z$.

A function which satisfies condition F_3, that each element of B is the image of some element of A under f, is said to be a *surjective* function or a *surjection*. Some authors prefer to describe such a function by the statement 'f is a function from A onto B' or 'f maps A onto B' and to use the notation $f : A \xrightarrow{\text{onto}} B$ to indicate this.

A function which satisfies condition F_4, that each element of the range of f is the image of precisely one element of the domain of f, is said to be an *injective* function or an *injection*. Such a function is sometimes described by the statement 'f is one-to-one' or 'f is 1–1', and indicated by a notation such as $f : A \xrightarrow{1-1} B$. A function which is not one-to-one is sometimes said to be many-to-one.

A function which satisfies both conditions F_3 and F_4 is said to be a *bijective* function or a *bijection*, or a one-to-one correspondence. The two conditions are illustrated in Fig. 3.7.

The inverse relation f^{-1} of a function f from a set A to a set B need not be a function. Since $f^{-1} \subseteq B \times A$, it will be a function if and only if

(i) $\mathscr{D}(f^{-1}) = B$;

(ii) if $(y, x) \in f^{-1}$ and $(y, z) \in f^{-1}$ then $x = z$.

For all relations f we have $\mathscr{D}(f^{-1}) = \mathscr{R}(f)$, so that condition (i) is precisely the condition F_3. Moreover, since $(y, x) \in f^{-1}$ stands

$f: A \to B$

A

B

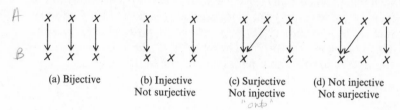

(a) Bijective   (b) Injective   (c) Surjective   (d) Not injective
    Not surjective   Not injective   Not surjective

"onto"

Figure 3.7

for $(x, y) \in f$, the condition (ii) is precisely the condition F_4. Thus the inverse relation f^{-1} is a function if and only if f is a bijection. In this case f^{-1} is also a bijection, and $f \circ f^{-1} = I_B$ and $f^{-1} \circ f = I_A$.

Even when a function f is not a bijection the inverse relation f^{-1} can be used to display an important property of the function f. Every relation S from a set A to a set B gives rise to a collection of subsets of A, one for each element of the range of S. The subset of A associated with the element $y \in \mathscr{R}(S)$ is $\{x \mid (x, y) \in S\}$. We shall here denote this subset by $S^{-1}(y)$. If S is a function, then the collection of subsets $C_S = \{S^{-1}(y) \mid y \in \mathscr{R}(S)\}$ is a partition of A. The proof of this statement makes use of the tautology of proof by parts and consists of showing three things:

(i) if S satisfies F_1 and F_2 then C_S satisfies P_1;

(ii) if S satisfies F_1 and F_2 then C_S satisfies P_2;

(iii) if S satisfies F_1 and F_2 then C_S satisfies P_3.

The proof of each of these statements is left as an exercise to the reader.

Now suppose that f is a function from A to B and that T is the equivalence relation on A associated with the partition C_f of A. Then $(x_1, x_2) \in T$ if and only if $f(x_1) = f(x_2)$. The elements of the quotient set A/T are the equivalence classes $[x] = f^{-1}(f(x))$. We leave as further exercises for the reader the proofs of the following statements. The relation $g = \{(x, X) \mid X = [x]\}$ is a surjective function. The relation $h = \{([x], y) \mid y = f(x)\}$ is an injective function such that $f = h \circ g$ and $\mathscr{R}(h) = \mathscr{R}(f)$. If f is a surjection, then h is also a surjection and is therefore a bijection. These statements are commonly displayed in a diagram such as Fig. 3.8.

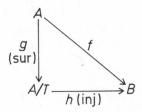

Figure 3.8

The function f is said to be factored into the surjection g and the injection h. The process of factoring general types of functions into special types of functions is found to be particularly valuable in algebra and in the more advanced stages of calculus.

Exercises 3.7

1. Show that the relation S defined from A to B is a function in each of the cases described below. Decide in each case whether the function is an injection, a surjection, a bijection, or none of these.

(a) $A = R \times R, B = R$.
   $((x, y), z) \in S$ if and only if $z = x^2 + y^2$.
(b) $A = R \times R, B = C$, the set of complex numbers.
   $((x, y), z) \in S$ if and only if $z = x + iy$.
(c) $A = R, B = R \times R$.
   $((t, (x, y)) \in S$ if and only if $x = t^2$ and $y = 2t$.
(d) $A = Z \times N, B = Q$.
   $((x, y), z) \in S$ if and only if $z = x/y$.
(e) $A = B = P$, the set of polynomials with real coefficients.
   $(p(x), q(x)) \in S$ if and only if $q(x) = d(p(x))/dx$.
(f) $A = P, B = R$.
   $(p(x), y) \in S$ if and only if $y = \int_0^1 p(x)\,dx$.
(g) $A = M$, the set of square real matrices, $B = R$.
   $(T, x) \in S$ if and only if $x = \det T$.
(h) $A = \mathscr{P}(U) \times \mathscr{P}(U), B = \mathscr{P}(U)$.
   $((X, Y), Z) \in S$ if and only if $Z = X \cup Y$.

2. Show that if $f: A \rightsquigarrow B$ and $g: B \rightsquigarrow C$ are bijections, then $g \circ f: A \rightsquigarrow C$ is a bijection and $(g \circ f)^{-1} = f^{-1} \circ g^{-1}$.

3. Show that if $f: A \rightsquigarrow B$ and $g: B \rightsquigarrow A$ have the properties that $f \circ g = I_B$ and $g \circ f = I_A$ then f is a bijection and $f^{-1} = g$.

4. Suppose that the domain and range of f are sets of real numbers. Then f is said to be strictly increasing if and only if

$$\forall x \in \mathscr{D}(f), \forall y \in \mathscr{D}(f), x < y \Rightarrow f(x) < f(y).$$

Show that if f is strictly increasing then f is a bijection. Is the converse of this result true?

5. Let A be the set of real valued functions with domain D. Can you use the relation $\leqslant$ in R to define a partial order relation in A? Is A a lattice with your partial ordering (see exercise 4 of §3.5)? Can you do this if the range of the functions in question is not a set of real numbers, but some other partially ordered set?

6. Consider the collection of all subsets of the set $\{a, b, c\}$, partially ordered by inclusion. This has the branching diagram of Fig. 3.9(a).

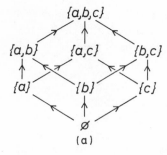 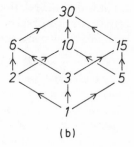

(a)

(b)

Figure 3.9

Consider the set of factors of 30 partially ordered by divisibility. This has the branching diagram of Fig. 3.9(b). Apart from the names of the elements these two diagrams are the same. We can find a bijection between the sets which preserves the partial ordering (formulate this symbolically). Show that the set of factors of 231 has the same diagram as these two.

Can you find a set of the second type which has a diagram like that of the collection of subsets of $\{a, b, c, d\}$ partially ordered by inclusion? Can you find a set of the first type which has a diagram like that of the set of factors of 12, partially ordered by divisibility?

3.8   FINITE AND INFINITE SETS

We have used the set of natural numbers and the set of real numbers in our examples of sets, relations and functions. Yet none of the theory of sets so far developed depends on a knowledge of these sets of numbers. Indeed, as was shown by mathematicians in the nineteenth century, the numbers can be defined within the theory of sets. There are many aspects to their work and only a few of them can be reviewed here.

First we consider an equivalence relation defined by means of bijective functions. Let $\mathscr{A}$ be a collection of sets and let S be the relation defined on $\mathscr{A}$ as follows,

$$S = \{(X, Y) \mid \text{there is a bijection from } X \text{ to } Y\}.$$

Since the identity function is a bijection the relation S is reflexive. Moreover, since the inverse of a bijection is a bijection and the composition of two bijections is a bijection, the relation S is symmetric and transitive. Hence S is an equivalence relation. In the remainder of this section, two sets will be said to be *equivalent* if they belong to the same equivalence class of this relation.

These equivalence classes were used by Cantor in the nineteenth century as the basis for a theory of the natural numbers and of transfinite numbers. The details of the development of the natural numbers from these equivalence classes alone are long and delicate. We shall by-pass them by characterizing the natural numbers as a set with properties which formalize the notion that 2 follows 1, 3 follows 2, 4 follows 3, and so on without end. The properties are referred to as the Peano axioms. Here the symbol A denotes a set.

Peano Axiom 1: There exists a function $s: A \rightsquigarrow A$ called the successor function.

Peano Axiom 2: There exists $e \in A$ such that $e \notin \mathscr{R}(s)$.

Peano Axiom 3: If $s(a) = s(b)$ then $a = b$, i.e. s is injective.

Peano Axiom 4: If $B \subseteq A$, $e \in B$ and $\forall x, x \in B \Rightarrow s(x) \in B$, then $B = A$.

Certainly the natural numbers as we understand them intuitively are seen to satisfy these axioms if we take $x+1$ for $s(x)$ and 1 for e. Moreover, these four axioms can be used to define processes of addition and multiplication in the set A which behave like the ordinary processes of arithmetic. It can be shown that if two sets A_1 and A_2 each satisfy the Peano axioms then they are equivalent, and moreover a bijection can be found which identifies the two successor functions. Thus we could apply the name 'the natural numbers' ambiguously to any set which satisfies the Peano axioms. Indeed, we do just this when we allow different names and symbols for the natural numbers. For example,

i,	ii,	iii,	iv,	...,	x,	xi,	...;
1,	2,	3,	4,	...,	10,	11,	...;
2^0,	2^1,	2^1+2^0	2^2,	...,	2^3+2^1,	$2^3+2^1+2^0$,	...;
un,	deux,	trois,	quatre,	...,	dix,	onze,	...;

are the beginnings of four ways of setting out a sequence of natural numbers.

The set of natural numbers has the property that it has many proper subsets to which it is equivalent. For example, if B denotes the set of even numbers, then $f: N \rightsquigarrow B, f: x \rightsquigarrow 2x$, is a bijection. On the other hand for any given natural number n the subset $\{1, 2, 3, ..., n\}$ of N does not have this property. If we express this property in the form 'there exists a subset of S which is equivalent to S and which is not equal to S', then we see that one form of its negation is 'every subset of S which is equivalent to S is equal to S', which can be expressed as 'the only subset of S which is equivalent to S is the set S itself'.

This property and its negation are used to characterize infinite and finite sets. A set S is said to be *finite* if the only subset of S which is equivalent to S is the set S itself. It can be shown that if S is finite then there is a natural number n such that S is equivalent to the set $\{1, 2, 3, ..., n\}$.

A set S is said to be *infinite* if it is equivalent to a proper subset of itself. Certainly the set N of natural numbers is infinite, but not all infinite sets are equivalent to N. For example, the set R_1 of real numbers between 0 and 1 is infinite and is not equivalent to N. We prove this statement by showing that the assumption that there is a bijection $f: N \rightsquigarrow R_1$ leads to the contradiction that there are two numbers which are equal but which are not equal, so by the tautology *reductio ad absurdum* there is no such bijection. The contradiction is established as follows. Assume that there is a bijection $f: N \leadsto R_1$. Let the image $f(n)$ of the natural number n be written in the decimal form $0 \cdot a_{1n} a_{2n} a_{3n} \ldots$, (decimals consisting of repeated 9's after a certain point are not allowed). For each n choose $b_n \in N$ such that $0 \leqslant b_n \leqslant 8$ and $b_n \neq a_{nn}$. Then $b = 0 \cdot b_1 b_2 b_3 \ldots \in R_1$, so that by the assumption there is a unique natural number m such that $b = f(m) = 0 \cdot a_{1m} a_{2m} a_{3m} \ldots$. However, since $b_m \neq a_{mm}$, it follows that $0 \cdot b_1 b_2 b_3 \ldots \neq 0 \cdot a_{1m} a_{2m} a_{3m} \ldots$, i.e. $b \neq f(m)$. The contradiction proves that the set of real numbers between 0 and 1 is not equivalent to the set of natural numbers.

On the other hand, the set of real numbers between 0 and 1 is equivalent to the set of real numbers between 0 and 2, an appropriate bijection being defined by the equation $y = 2x$. Moreover, the set of real numbers between 0 and 1 is equivalent to the set of all real numbers. These sets are infinite sets, and are in a sense infinitely larger than the set of natural numbers. It can be shown that in this same sense the power set $\mathscr{P}(S)$ of an infinite set S is infinitely larger than S itself, and using this fact we can generate a sequence of sets

$$S, \mathscr{P}(S), \mathscr{P}(\mathscr{P}(S)), \mathscr{P}(\mathscr{P}(\mathscr{P}(S))), \ldots,$$

each infinitely larger than the one before.

The further study of these results is fascinating, but we can do no more at this stage than give some examples of sets which are equivalent to the set of natural numbers. Such sets are said to be *denumerably infinite* or *countably infinite*. Sets which are either finite or countably infinite are said to be *countable* or *denumerable*, while sets which are not countable are said to be *uncountable* or *nondenumerable*.

For the first example a bijection can be defined explicitly. Intuitively, the set Z of integers can be set out in the order

$$0, 1, -1, 2, -2, 3, -3, \ldots.$$

Formally, there is a bijection

$$f: N \leadsto Z, \quad f: n \leadsto \begin{cases} n/2, & n \text{ even}, \\ -(n-1)/2, & n \text{ odd}. \end{cases}$$

Alternatively,

$$f = \{(2n,n)\,|\,n \in N\} \cup \{(2n-1, -n+1)\,|\,n \in N\}.$$

This function shows that Z is denumerably infinite.

For the second example, it is difficult to write out explicitly the equations for a bijection but easy to describe a process by which one can be obtained. We show that the set of positive rational numbers is denumerably infinite. The numbers are set out in rows keeping those with a given denominator in a row.

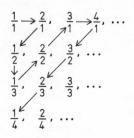

The numbers are then written out in the order indicated by the arrows, ignoring those which have already been included in some other fractional form, and the natural numbers are written in order underneath.

1,	2,	$\frac{1}{2}$,	$\frac{1}{3}$,	3,	4,	$\frac{3}{2}$,	$\frac{2}{3}$,	$\frac{1}{4}$,	$\frac{1}{5}$,	5,	6,	$\frac{5}{2}$, ...
↑	↑	↑	↑	↑	↑	↑	↑	↑	↑	↑	↑	↑
1,	2,	3,	4,	5,	6,	7,	8,	9,	10,	11,	12,	13, ...

A bijection is obtained by associating with each natural number in the lower line the rational number above it in the upper line.

The elements of a denumerably infinite set may be statements. For example, the binomial theorem for positive integer indices can be written

$$\forall n \in N, (1+a)^n = 1+na+ \ldots + \frac{n!}{r!(n-r)!}\,a^r + \ldots + a^n.$$

This theorem and other theorems which involve a denumerably infinite set of statements are commonly proved 'by induction'.

If the theorem to be proved is $\forall n \in N$, $P(n)$, then the process of proof is to prove that $P(1)$ is true, and that for all $k \in N$, if $P(k)$ is true then $P(k+1)$ is true. It then follows from the Peano axioms that the theorem is true. For if S is the set $\{n \in N \mid P(n)\}$ the process of proof ensures that $1 \in S$ and that for all $k \in N$, if $k \in S$ then $k+1 \in S$. Thus by Peano axiom 4 (with N for A and S for B) we have $S = N$. For ease of reference, this axiom is often called the *Axiom of Induction*. We give an example of proof by induction.

Proposition 3.1 For every natural number n, $2^n \geqslant 1 + 2n$.

Proof Let $S = \{n \in N \mid 2^n \geqslant 1+2n\}$. Since $2^1 = 1+1$, we have $1 \in S$. Now suppose that $k \in S$, i.e. that $2^k \geqslant 1+2k$. Then

$$2^{k+1} = 2 . 2^k \geqslant 2(1 + 2k) = 2+4k$$
$$= 2+2k+2k \geqslant 1+2k+2 \quad \text{(since } k \geqslant 1\text{)}.$$

Hence

$$2^{k+1} \geqslant 1+2(k+1), \qquad \text{i.e. } k+1 \in S.$$

Thus by the axiom of induction $S = N$.

Exercises 3.8

1. Generalize the discussion of the denumerability of the set of rational numbers to show that the union of a denumerable collection of denumerable sets is itself a denumerable set.

2. Show that the sets $A = \{x \mid x \in R \text{ and } a < x < b\}$ and $B = \{x \mid x \in R \text{ and } c < x < d\}$ are equivalent.

3. Show that the sets $A = \{x \mid x \in R \text{ and } 0 < x < 1\}$ and $B = \{x \mid x \in R \text{ and } 0 \leqslant x < 1\}$ are equivalent.

4. Show that if A is nondenumerable and B is denumerable, then $A \backslash B$ is nondenumerable. Deduce that the set of irrational numbers between 0 and 1 is nondenumerable.

5. Two million points are given in the coordinate plane. Is it possible to find a line with the property that a million of the points lie on either side of this line?

6. Prove by induction that for all natural numbers n,

$$(A_1 \cup \ldots \cup A_n) \cap B = (A_1 \cap B) \cup \ldots \cup (A_n \cap B).$$

7. Discuss the following proof by induction that all natural numbers are equal.

Let $P(n)$ stand for the statement 'Given any set of n natural numbers, the numbers are all equal'. Clearly $P(1)$ is true, since a number is equal to itself. Let us assume that $P(k)$ is true. Let $a_1, \ldots, a_{k+1}$ be a set of $k+1$ natural numbers. By hypothesis $P(k)$ is true and so $a_1, \ldots, a_k$ are all equal, and also $a_2, \ldots, a_{k+1}$ are all equal. Hence $a_1, \ldots, a_{k+1}$ are all equal, and so $P(k)$ implies $P(k+1)$. Thus, by induction, all natural numbers are equal.

8. Suppose that $a_1 = 0$, $a_2 = 3$, $a_3 = 2$, and $a_n = a_{n-1} + a_{n-2} - a_{n-3}$. Find a general formula for a_n in terms of n, and prove your result by induction.

9. Discuss the possibility of proof by induction of the following results.

(a) $\sum_{r=1}^{n} r = \frac{1}{2}n(n+1)$.

(b) $\sum_{r=1}^{n} (2r-1) = n^2$.

(c) $\sum_{r=1}^{n} r^2 = (2n^2+1)(2n+2)/12$.

(d) $n^2 - n + 41$ is prime.

(e) $(1+a)^n \geqslant 1 + na$ if $a \geqslant -1$.

4

MATHEMATICAL THEORIES

The development of mathematics can be viewed as the interaction of two processes. The first is the identification of characteristics which are common to large collections of objects and situations. The second is the description of differences between objects and situations. The work of the previous chapter illustrates these two processes. We first of all isolated the idea of a relation which was common to many situations. Then we looked at equivalence relations, partial order relations and functions. These were different kinds of relations, yet each in itself was applicable to many situations both inside and outside mathematics. Similarly, within the idea of a function we distinguished between injective, surjective and bijective functions.

In this chapter we shall see how these two processes influence the further development of the language commonly used for some important mathematical models. Firstly, functions which have for their range a set X and for their domain a set $X \times X$ are considered. Special types of such functions lead to the idea of a group, which is in itself a general model for many physical situations. Subsequently, functions with domain $X \times X$ and range R are used to develop the idea of a metric space. This will be used as a model for a variety of situations in which distance is measured. Finally, we shall introduce the notion of a measure space. It will arise from the consideration of the measurement of certain quantities such as mass and area, but in the next chapter it will prove to be applicable also to the discussion of probabilities.

4.2 BINARY OPERATIONS

The mathematical process to be studied in this section is that of combining two mathematical objects to form a third object of the same kind. We shall consider a number of situations where this occurs, and use the language developed in previous chapters to build a conceptual model which reflects aspects shared by these situations. The model will then be used to describe differences between them.

Examples of this process at work have already occurred in this book. For example, in Chapter 1 a number of ways in which two sets can be combined to form a third set were described. In Chapter 2, we did the same with statements, developing the logical connectives to that end; while in Chapter 3 some situations were explored in which functions could be combined. Table 4.1 contains some of these, together with familiar examples involving numbers. We shall use these examples as a basis for the model. In the table, the two objects to be combined are printed in the second and third columns, and the symbol for the combination is printed in the fourth column. The rows are numbered for later reference.

TABLE 4.1

1	$A \subseteq U$	$B \subseteq U$	$A \cup B$
2	$A \subseteq U$	$B \subseteq U$	$A \cap B$
3	$A \subseteq U$	$B \subseteq U$	$A \backslash B$
4	$A \subseteq U$	$B \subseteq U$	$A \triangle B$
5	$g: A \rightsquigarrow A$	$h: A \rightsquigarrow A$	$h \circ g: A \rightsquigarrow A$
6	$x \in N$	$y \in N$	$x + y$
7	$x \in R$	$y \in R$	$x + y$
8	$x \in R$	$y \in R$	$x - y$
9	$x \in Q$	$y \in Q$	$x \cdot y$
10	$x \in Q \backslash \{0\}$	$y \in Q \backslash \{0\}$	$x \div y$
11	$x \in N$	$y \in N$	x^y
12	$x \in R$	$y \in R$	$\max(x, y)$

You may have met similar situations involving matrices, permutations of an ordered collection of objects, clock arithmetic, rotations of a circle, symmetries of a polygon, translations in the plane, etc. You should extend Table 4.1 with as many other situations as you can, and consider the extended table alongside the following discussion, examining each example in relation to the properties discussed.

The first decision to be made is whether the circumstances are such that the order in which two elements are combined is important. A study of the table will show that in some examples the order of elements is important, whilst in others it is not. For example, in general $A\backslash B \neq B\backslash A$ and $x-y \neq y-x$, whilst it is always true that $A\cup B = B\cup A$ and $x.y = y.x$. Since order is important in some of the situations we shall consider combinations of elements of an ordered pair, and later on distinguish those situations where order is not relevant.

In each of the examples in the table, with a pair of elements of some set X there is associated another element of X. In the first five examples the set X corresponds to the power set $\mathscr{P}(U)$, and in the last seven examples the set X is an appropriate set of numbers. In example 5, the set X can be taken to be the set of all functions whose domain and range is the set A.

Since we have decided upon using ordered pairs, to each example in the table there corresponds a set X, and with each element of $X \times X$ there is associated a unique element of X; in other words there is a function from $X \times X$ to X. It will be convenient to introduce notations for the functions defined in this way. For the remainder of this section f_1 will be used to correspond to example 1 of the table, f_2 to example 2, etc.

Now f_7 is a subset of $(R \times R) \times R$. In the rule method of writing sets

$$f_7 = \{((x,y), z) \,|\, x, y, z \in R \text{ and } x+y = z\}.$$

It was pointed out in §3.7 that an open statement such as $y=x^2$ which defines a function is not itself a function, even though one may commonly speak of 'the function $y=x^2$', leaving it to the reader to decide on a suitable domain. The properties of the function depend on the domain as well as on the open statement $y=x^2$. Similarly, as we shall see, the properties of the operation of addition depend not only upon the open statement $x+y=z$ but also on the universes of discourse for the variables. Thus while one commonly speaks of addition as an operation without reference to the domain involved, here the term *operation* of addition will be used for functions such as f_6 defined by addition on specific domains. The other functions corresponding to the examples in the table will also be called operations, and since the operations associate with two elements of X a single element of X, they will be called binary operations. Setting out the special form of the definition of a function in this case, we have the following definition.

H

Definition 4.1   A *binary operation* f on a set X is a relation $f \subseteq (X \times X) \times X$ such that

$F_1 : \mathscr{D}(f) = X \times X$

$F_2 :$ if $((x, y), z_1) \in f$ and $((x, y), z_2) \in f$ then $z_1 = z_2$.

Let f be a binary operation on a set X. Then a subset Y of X is said to be *closed under* f if and only if $f(Y \times Y) \subseteq Y$, i.e.

$$\forall x, y \in Y, \quad f((x, y)) \in Y.$$

In this case $f \cap ((Y \times Y) \times Y)$ is a binary operation on Y. We sometimes say that a set Y is closed under a binary operation without specifying the set X on which the binary operation is defined. For example, we may say that the set $\{x \mid x \in R$ and $0 < x < 1\}$ is closed under multiplication and is not closed under addition, leaving it for the reader to fill in 'the real numbers' or 'the positive real numbers' as a set X on which the binary operations are defined.

The use of the notations $x + y$, $A \cup B$, etc., for the images of pairs of elements under the action of certain binary operations is so familiar that it is helpful to devise similar notations for images under general binary operations. Connecting symbols such as $*$ and $\circ$ can be used, and a notation such as

$$f : X \times X \rightsquigarrow X, \qquad f : (x, y) \rightsquigarrow x*y.$$

Thus $x*y$ is a useful alternative notation for $f((x, y))$, and we shall sometimes speak of a binary operation $*$ on a set X. If Y is a subset of X which is closed under $*$ we shall speak of the binary operation $*$ on Y, using the same symbol $*$ in spite of the difference in domain. When using the functional notation, we shall abbreviate $f((x, y))$ to $f(x, y)$ to avoid duplication of parentheses.

Since a binary operation f on a set X is itself a set, being a subset of $(X \times X) \times X$, it may be convenient to specify it by the rule method or by the list method. The binary operations f_1 to f_{12} were specified in Table 4.1 by a method which corresponds to the rule method for sets. We shall now describe two binary operations by the list method. Let $A = \{a, b, c\}$, so that

$$A \times A = \{(a, a), (a, b), (a, c), (b, a), (b, b), (b, c), (c, a), (c, b), (c, c)\}.$$

To specify a function $f_{13} : A \times A \rightsquigarrow A$ it is necessary to associate with each one of these nine ordered pairs an element of A. If these elements are chosen in order as $a, c, b, c, a, c, b, c, b$, the function is $f_{13} = \{((a, a), a), \dots\}$ with the nine appropriate ordered pairs listed. The function can be represented concisely by means of a block diagram in which the elements of the set A are listed vertically and horizontally, so that each position in the diagram is identified by an element of $A \times A$, and the corresponding element of A is inserted in that position.

	a	b	c
a	a	c	b
b	c	a	c
c	b	c	b

Block diagram for f_{13}: $\{a, b, c\} \times \{a, b, c\} \rightsquigarrow \{a, b, c\}$

This diagrammatic method of representing binary operations is particularly useful when the set X is reasonably small and yet too large for it to be convenient to list all the elements of a function defined on $X \times X$. The next figure represents a binary operation f_{14} on the set $B = \{a, b, c, d, e\}$ consisting of five elements.

	a	b	c	d	e
a	c	a	d	e	c
b	a	b	c	d	e
c	d	c	b	b	a
d	e	d	b	d	d
e	c	e	b	d	b

Block diagram for f_{14}: $B \times B \rightsquigarrow B$

In the diagram for f_{14} the vertical listing of the set B represents the first component of the elements of $B \times B$, and the horizontal listing represents the second component. Thus to find $f_{14}(e, c)$ one looks at the entry in the row corresponding to e and in the column corresponding to c. That entry is b, and so $f_{14}(e, c) = b$.

Having expressed in functional terms a concept which unifies the various examples in the list, we now consider some of the auxiliary properties which serve to distinguish between binary operations.

Commutative Operations

As was remarked previously, we chose to define a binary operation as a function whose domain is a set of ordered pairs, although in several of the examples in Table 4.1 the order of the elements to be combined is not important. For example, for the union of two sets A and B it does not matter whether one writes $A \cup B$ or $B \cup A$: this was expressed in terms of formal proof in Proposition 2.2. Again, when dealing with numbers, the result of adding 3 to 5 is the same as the result of adding 5 to 3. The equations $A \cup B = B \cup A$, $3 + 5 = 5 + 3$, which represent the statements above, lead to a consideration of those operations $*$ for which $x*y = y*x$. When formulated functionally this leads to the following definition.

Definition 4.2   A binary operation $f : X \times X \rightsquigarrow X$ is said to be *commutative* if and only if

$$\forall x, y \in X, \quad f(x, y) = f(y, x).$$

Thus, as has already been seen, the operations f_1, f_6, f_7 are commutative. On the other hand the binary operation f_{11} is not commutative. To justify this latter statement it is necessary to show that $\sim(\forall x, y \in N, x^y = y^x)$ is true, i.e. that $\exists x, y \in N, x^y \neq y^x$ is true, which can be done by exhibiting the pair $(2, 3)$, for which $2^3 \neq 3^2$.

Associative Operations

We now consider ways of combining more than two objects. In order to form the intersection of three sets A, B and C on a Venn diagram, one simply picks out visually the part common to all three sets. Bearing in mind the way in which the concept of a binary operation was formulated, one may now invent the concept of a ternary operation as a function $f : X \times X \times X \rightsquigarrow X$. Similarly, one could have an n-ary operation for combining n objects. However, to take a ternary operation as a model for adding three numbers does not reflect the way in which one usually performs such an operation. For example, to perform the addition $12 + 9 + 13$, one might proceed as follows: $12 + 9 = 21$, $21 + 13 = 34$. Addition has here been considered as a binary operation which has been repeated. In the above example some people would add the numbers in the following way: $9+13 = 22$, $12+22 = 34$. The order of addition makes no difference to the final result. However, for some binary operations the situation is different. For example, if the binary operation f_{11} is applied to $2, 3$ and 2 in two different ways, one obtains $(2^3)^2 = 8^2 = 64$, while $(2)^{3^2} = 2^9 = 1024$. Again, with the operation f_8, one has $(8-4)-2 = 2$, whereas $8-(4-2) = 6$. These examples suggest that it may be useful to have a terminology to distinguish between those binary operations $*$ for which $(x*y)*z$ and $x*(y*z)$ are always equal and those for which they are sometimes unequal. In the following definition the condition is expressed in the functional notation.

Definition 4.3   A binary operation $f : X \times X \rightsquigarrow X$ is said to be *associative* if and only if

$$\forall x, y, z \in X, \; f(f(x, y), z) = f(x, f(y, z)).$$

Thus addition is an associative operation, and if multiple intersections are regarded as repeated binary operations, then intersection is also as associative operation.

Identity elements

If asked to perform the multiplication $3.1.4.\frac{1}{2}.1.1.2.1$, one would simplify this initially to $3.4.\frac{1}{2}.2$ before proceeding any further. Similarly, if in a calculation involving functions, as with the operation f_5, an expression of the form $(h \circ I_A) \circ (g \circ I_A)$ occurred, this would immediately be written as $h \circ g$. A situation where this kind of thing occurs will clearly be useful for calculations, and so it would seem worthwhile to isolate the relevant feature. In the alternative notation, one must look for elements e with the property that $x*e = x$ for all x. Bearing in mind that not all operations are commutative, one should also consider whether there are elements d such that $d*x = x$ for all x. In the example involving multiplication of real numbers the number 1 satisfies $1.x = x.1 = x$ for all real numbers x. In this case the operation is commutative. Composition of functions is not a commutative operation, but nevertheless, for f_5, one has $I_A \circ g = g \circ I_A$ for all functions $g : A \rightsquigarrow A$. The following definition is therefore introduced.

Definition 4.4 An element $e \in X$ is said to be an *identity element* for a binary operation $f: X \times X \rightsquigarrow X$ if and only if

$$\forall x \in X, \quad f(e, x) = f(x, e) = x.$$

The number 0 is an identity element for the operation f_7, but there is no identity element for the operation f_6. Similarly the empty set $\varnothing$ is an identity element for the operation f_1.

For the operation f_8, there is no element e with the property that $\forall x \in R$, $e - x = x - e = x$; however the number 0 satisfies $x - 0 = x$, but not $0 - x = x$. A similar situation occurs with f_{11}, where $\forall x \in N$, $x^1 = x$, but there is no element d satisfying $\forall y \in N$, $d^y = y$. These situations can be dealt with by the formulation of concepts of left and right identity elements, but we shall not do so here.

Inverse elements

In the first example involving identity elements, the multiplication was reduced from $3.1.4.\frac{1}{2}.1.1.2.1$ to $3.4.\frac{1}{2}.2$, because of the role of the number 1 as an identity. A further reduction can be made because $\frac{1}{2}.2 = 1$, and so $3.4.\frac{1}{2}.2 = 3.4.1 = 3.4$. In general, for every non-zero real number x, there is a real number y such that $x.y$ is the identity 1 for multiplication. On the other hand, as was noted in §3.7, for some functions $f: A \rightsquigarrow A$ there is a function g such that $g \circ f$ is the identity function I_A, and for some there is not. A concept which distinguishes between phenomena of these kinds is defined as follows.

Definition 4.5 Let f be a binary operation on a set X with a unique identity element e. Then, given an element $x \in X$, an element $y \in X$ such that $f(x, y) = f(y, x) = e$ will be called an *inverse element for x*.

Thus relative to the binary operation f_7 every element $x \in R$ has a unique inverse element $(-x)$. In contrast, there are no inverse elements for a non-empty set A under the binary operation f_1.

The notion of a binary operation, and the auxiliary notions of commutativity, associativity, identity elements and inverse elements are useful for the description of many mathematical processes, some of which have been exemplified above. It often happens that when one performs a calculation in such a situation, one does so by means of repeated use of the binary operation. However, this is not always so. As was pointed out in the introduction to associative operations, one sometimes works with a ternary operation model of intersection of sets. Similarly, many people occasionally work with a ternary operation model of addition when they recognize certain familiar number patterns such as $2+2+2$ or $1+2+3$. Indeed every associative binary operation gives rise to a ternary operation, and it is probably true that in familiar situations people work with a ternary operation model for ease and speed.

It is equally true that every commutative binary operation $f : X \times X \rightsquigarrow X$ gives rise to a family of unary operations

$$\{f_x : X \rightsquigarrow X, \quad f_x : y \rightsquigarrow f(x,y) \mid x \in X\}$$

with the property that $\forall x, y \in X$, $f_x(y) = f_y(x)$. Conversely, a family of unary operations $\{g_x : X \rightsquigarrow X \mid x \in X\}$ such that $\forall x, y \in X$, $g_x(y) = g_y(x)$ gives rise to a commutative binary operation $g : X \times X \rightsquigarrow X$ defined by $g(x, y) = g_x(y)$.

Just as there are occasions when people work with a ternary operation, so there are occasions when people work with the family of unary operations rather than the binary operation. For example, if a particular product of natural numbers will not come immediately to mind, many of us will recover it by working through an appropriate multiplication table. Similarly, for addition, some approaches to arithmetic at the primary level exhibit 'add four to three' by a notation such as $3 \xrightarrow{\text{add } 4} 7$. Here the '4' and the '3' do not seem to have the same conceptual status. It appears that 'add 4' is acting on '3' to give the result '7'. This can be represented by a function $a_4 : N \rightsquigarrow N$ such that $a_4(3) = 7$. In general, one has a family of functions for addition,

$$\{a_n : N \rightsquigarrow N, \quad a_n : m \rightsquigarrow m+n \mid n \in N\}.$$

From the point of view of general theory rather than special examples, such families of unary operations have two disadvantages. Firstly, a non-commutative binary operation cannot be replaced by a one-parameter family of unary operations; two families are needed, one to express action on the left and the other to express action on the right. Secondly, the equation, for a family of unary operations $\{f_x \,|\, x \in X\}$, corresponding to the equation which defines associativity of a commutative binary operation f involves double suffices:

$$\forall x, y, z \in X, \quad f_z(f_y(x)) = f_{f_z(y)}(x).$$

This is extremely difficult to remember and to read, and the corresponding condition for f is much better in these respects. It is even easier to read the condition when printed in the alternative notation for binary operations:

$$\forall x, y, z \in X, \quad (x*y)*z = x*(y*z).$$

Nevertheless, in unfamiliar and difficult situations it seems that most people like to have only one variable to deal with at a time, and turn to a family of unary operations as the model with which they will work. The problem of defining addition and multiplication within a set which satisfies the Peano axioms gives rise to such a situation. In §3.8 we discussed inductive proofs of statements of the form $\forall n \in N, P(n)$. A similar process can be used to define inductively a family of functions $\{s_n \,|\, n \in N \text{ and } s_n : N \rightsquigarrow N\}$ such that the operation s_n takes each natural number r into the nth successor after r. The family can be defined formally as follows in terms of the successor function $s : N \rightsquigarrow N$ specified in the Peano axioms. Let $s_1 = s$, and if s_n is defined let $s_{s(n)} : N \rightsquigarrow N$, be given by $s_{s(n)} : r \rightsquigarrow s(s_n(r))$. It then follows from Peano axiom 4 that $\{s_n\}$ is a well defined family of functions which has all the properties that one expects of a unary operations model of addition of the natural numbers. A second family of unary operations can be defined in terms of the family $\{s_n\}$ in such a way as to have all the properties that one expects of a unary operations model of multiplication of the natural numbers. The details of the proofs of these statements are too long to be included here. There are many steps, each of which involves the use of Peano axiom 4. It is sufficient to note that the process gives rise to unary operation models of the arithmetical operations. By a change of notation one can obtain binary operation models of addition and multiplication. These are commonly used for calculations which are neither so easy that a ternary (or other n-ary) model is more appropriate, nor so hard as to make recourse to the unary model necessary.

Exercises 4.2

1.(a) Is the set of rectangular regions in R^2 closed under f_1?
  (b) Is the set of rectangular regions in R^2 closed under f_2?
  (c) Is the set of polygonal regions in R^2 closed under f_3?
  (d) Is the set of even natural numbers closed under f_6?
  (e) Is the set of odd natural numbers closed under f_6?
  (f) Is the set of irrational numbers closed under f_7?
  (g) Is the set Z closed under f_{10}?
  (h) Is the set of odd natural numbers closed under f_{11}?
  (i) Is the set $\{a, c, e\}$ closed under f_{14}?

2. For each of the binary operations f_1 to f_{14}, decide whether it is commutative, whether it is associative, whether an identity element exists, and whether inverse elements exist.

3. Describe by the rule method the binary operation defined by the diagram below, and investigate its properties.

	a_0	a_1	a_2	a_3
a_0	a_0	a_1	a_2	a_3
a_1	a_1	a_2	a_3	a_0
a_2	a_2	a_3	a_0	a_1
a_3	a_3	a_0	a_1	a_2

4. Prove that if a binary operation has an identity element, then it is unique.

5. Find a binary operation f on a set A such that some element in A has more than one inverse element.

6. If $*$ and $\circ$ are two binary operations defined on a set X, the operation $*$ is said to be distributive over the operation $\circ$ if and only if

$$\forall x, y, z \in X, \quad x*(y \circ z) = (x*y) \circ (x*z).$$

Investigate distributivity between appropriate pairs of binary operations from Table 4.1.

4.3 GROUPS

We have isolated certain concepts which serve to distinguish various properties of binary operations, and have seen that several of our examples satisfy more than one of the properties. For example, the binary operation f_1 of set union is commutative, associative and has an identity element $\varnothing$. The operation f_7 of addition of real numbers is commutative, associative and has an identity element and inverse elements. The operation f_5 of composition of functions from a set A to itself is associative but not commutative: it has an identity element, and each bijective function has an inverse element. Readers who have worked with 2×2 matrices will have found that multiplication is an associative operation with an identity element, and that a certain type of matrix always has an inverse, but that it is not a commutative operation. In these and many other situations in mathematics the same combination of properties occurs sufficiently often to make it worthwhile to give special attention to such situations. Many physical phenomena such as motions of elementary particles, and crystal structure, when modelled by the mathematical concept of a binary operation, have properties which give rise to this same combination of mathematical properties within the model, and this gives an added reason for studying this particular structure. In this section the notion of a group will be introduced as a mathematical model for these situations. We shall investigate some properties of groups and make comments on the processes of proof in this context, as in §2.7. The detailed mathematical work will be more involved than in that section however, and we shall not relate each step to a tautology, but make remarks of a more general nature about the proofs.

Definition 4.6 A *group* is an ordered pair (G, f) where G is a set and f is an associative binary operation on G with an identity element such that each element of G has an inverse.

In considering exercise 4 of §4.2 you should have proved that for every binary operation with an identity element, that identity is unique. We shall here use the notation e for the unique identity of a group (G, f) and shall denote an inverse of an element x by x'. Where it is convenient to do so, we shall use the alternative notation $x*y$ for $f(x, y)$. In this notation, a group is an ordered pair $(G, *)$ where G is a set on which is defined a binary operation $*$ such that

$$G_1 : \forall x, y, z \in G, \quad (x*y)*z = x*(y*z);$$

$$G_2 : \exists e \in G, \quad \forall x \in G, \quad x * e = e * x = x;$$

$$G_3 : \forall x \in G, \quad \exists x' \in G, \quad x * x' = x' * x = e.$$

If the binary operation $*$ is commutative, then the group $(G, *)$ will be said to be commutative or abelian.

Before considering consequences of this definition, we shall look at some examples of groups. Let A be a set and let G be the set of bijections from A to itself. The identity function $I_A : A \rightsquigarrow A$, $I_A : x \rightsquigarrow x$, is certainly a bijection, and is an identity element for the binary operation of composition of functions. It follows from the work and examples of §3.7 that if $f, g \in G$ then $f^{-1} \in G$, $g \circ f \in G$ and $f \circ f^{-1} = f^{-1} \circ f = I_A$. Moreover, composition of functions is associative. Hence $(G, \circ)$ is a group.

The set of complex numbers with unit modulus with the binary operation of multiplication forms a group: the identity element of the group is the complex number 1, the inverse of z is $\bar{z}$, the conjugate of z, and the operation is associative. The set of polynomials of degree less than three with the binary operation of polynomial addition forms a group. The set of 2×2 matrices of unit determinant with the binary operation of matrix multiplication forms a group.

We now turn to some of the consequences of the definition of a group. While the identity element of a binary operation must be unique, if it exists at all, you were asked to show in exercise 5 of §4.2 that inverse elements need not be unique. The first result which we shall prove here is that for an associative operation inverse elements are unique.

Proposition 4.1   If $(G, *)$ is a group then every element $x \in G$ has a unique inverse element.

Discussion of Proof   We show that if two symbols for elements of G satisfy the condition for an inverse of x then the two symbols stand for the same element. The condition G_3 guarantees the existence of an inverse element x'. We suppose that an element y also has the property that $x * y = y * x = e$. The result that $y = x'$ will depend on the associativity of the binary operation, applied to three suitably chosen elements. The proof is constructed by trial and error moving in two directions from the expression $y * (x * x')$, but it is presented in a different order.

Proof By condition G_3, there exists an element x' inverse to x. Suppose also that $y*x = x*y = e$. Then

$$\begin{aligned} y &= y*e && \text{(by } G_2) \\ &= y*(x*x') && \text{(by } G_3) \\ &= (y*x)*x' && \text{(by } G_1) \\ &= e*x' && \text{(by assumption)} \\ &= x' && \text{(by } G_2). \end{aligned}$$

Thus the element x' inverse to x is unique.

It follows immediately from this result that for each element $x \in G$ we have $(x')' = x$, for $(x')'$ is an inverse of x' and also x is an inverse of x', whence by uniqueness of inverses they are the same element.

We can now prove a result which gives more information about inverse elements.

Proposition 4.2 If $(G, *)$ is a group and $x, y \in G$ then $(x*y)' = y'*x'$.

Discussion of Proof The notation $(x*y)'$ stands for the inverse of $x*y$ which we know to exist by G_3, and to be unique by the previous proposition. Thus if $y'*x'$ is an inverse of $x*y$, it must be an alternative notation for the element $(x*y)'$. To show that $y'*x'$ is an inverse of $x*y$ we have to show that $(x*y)*(y'*x') = e$ and that $(y'*x')*(x*y) = e$. The details of the manipulation of each of these are so similar that only one of them is written out in full in the following proof.

Proof

$$\begin{aligned} (x*y)*(y'*x') &= x*(y*(y'*x')) && \text{(by } G_1) \\ &= x*((y*y')*x') && \text{(by } G_1) \\ &= x*(e*x') && \text{(by } G_3) \\ &= x*x' && \text{(by } G_2) \\ &= e && \text{(by } G_3). \end{aligned}$$

Similarly $(y'*x')*(x*y) = e$. Hence $y'*x'$ is an inverse of $x*y$. Since, by Proposition 4.1, inverses are unique, $y'*x' = (x*y)'$.

Further developments of the theory of groups depend on the introduction of auxiliary concepts which are inspired by particular applications of the theory. A proposition will preceed the definition of the first auxiliary concept, that of a subgroup.

Proposition 4.3   Let $(G, *)$ be a group and let H be a subset of G with the following properties:

$$S_1 : e \in H;$$

$$S_2 : \forall x \in H, \quad x' \in H;$$

$$S_3 : \forall x, y \in H, \quad x * y \in H.$$

Then $(H, *)$ is a group.

Proof   By condition S_3, the set H is closed under the binary operation $*$ on G, and by the convention introduced in the last section the same symbol $*$ will be used for the corresponding binary operation on H. Since the operation $*$ on G is associative, so is the operation $*$ on H. Property S_1 guarantees that the identity element for the operation $*$ on G belongs to H and so is an identity element for the operation $*$ on H. Property S_2 asserts that for each element x of H, its inverse element x' with respect to the operation $*$ on G belongs to H, and so is an inverse element for x with respect to the operation $*$ on H. Thus $(H, *)$ is a group.

Definition 4.7   Let $(G, *)$ be a group and let H be a subset of G which satisfies the conditions S_1, S_2 and S_3. Then the group $(H, *)$ is said to be a *subgroup* of $(G, *)$.

Given a subgroup $(H, *)$ of a group $(G, *)$ we can define various relations on G and prove results such as the following.

Proposition 4.4   If $(H, *)$ is a subgroup of $(G, *)$, then the relation

$$S = \{(x, y) \mid x, y \in G \text{ and } y * x' \in H\}$$

is an equivalence relation.

Proof   (i) If $x \in G$ then

$$x * x' = e \qquad (\text{by } G_3).$$

Hence

$$x*x' \in H \qquad \text{(by } S_1 \text{)}.$$

Thus the relation S is reflexive.

(ii) If $(x, y) \in S$ then

$$y*x' \in H \qquad \text{(by definition)}.$$

Hence

$$(y*x')' \in H \qquad \text{(by } S_2 \text{)}$$

i.e.

$$x*y' \in H \qquad \text{(by Proposition 4.2)}$$

so that

$$(y, x) \in S.$$

Thus the relation S is symmetric.

(iii) If $(x, y) \in S$ and $(y, z) \in S$ then

$$y*x' \in H \text{ and } z*y' \in H \qquad \text{(by definition)}.$$

Thus

$$(z*y')*(y*x') \in H \qquad \text{(by } S_3 \text{)}.$$

The proof of Proposition 4.2 can be modified to show that

$$(z*y')*(y*x') = z*x'.$$

Hence $z*x' \in H$, and so $(x, z) \in S$. Thus the relation is transitive.

It follows by proof by parts that S is an equivalence relation.

As was shown in §3.4, to every equivalence relation S on a set G there corresponds a partition of G. In this case, G is partitioned into disjoint sets

$$H(x) = \{y \,|\, (x, y) \in S\} = \{y \,|\, y*x' \in H\},$$

which are called the *right cosets* of the subgroup $(H, *)$ in $(G, *)$. In this case the standard notation G/S for the quotient set of the equivalence relation is replaced by G/H. Note that $\forall x \in G,\ x \in H(x)$. Note also that $H(x_1)$ and $H(x_2)$ may denote the same set even when

x_1 and x_2 are different. A condition for two cosets to be equal is given in the following proposition.

Proposition 4.5   Two right cosets $H(x_1)$ and $H(x_2)$ are equal if and only if $x_1 * x_2' \in H$ and $x_2 * x_1' \in H$.

Discussion of Proof   With this form of wording the statement $H(x_1) = H(x_2)$ can be regarded as the conclusion of the proposition and the statement $(x_1 * x_2' \in H) \wedge (x_2 * x_1' \in H)$ can be regarded as the condition. The proof is presented in two parts. One of these consists of a proof that the condition implies the conclusion, ending with the words 'thus the condition is sufficient'. This deals with the part of the proposition 'the conclusion holds if the condition holds'. The other comprises a proof that the conclusion implies the condition, ending with the words 'thus the condition is necessary'. This deals with the part of the proposition 'the conclusion holds only if the condition holds'. It does not matter in which order these two parts are dealt with, and in fact we look at the latter first.

(i) We prove that if $H(x_1) = H(x_2)$ then $x_1 * x_2' \in H$, claim that $x_2 * x_1' \in H$ follows in a similar way, and that the required result follows using the tautology of proof by parts.

(ii) The conclusion $H(x_1) = H(x_2)$ holds if the following statement is true:

$$(\forall y \in H(x_1), y \in H(x_2)) \wedge (\forall y \in H(x_2), y \in H(x_1)).$$

We prove that if $x_1 * x_2' \in H$ then $\forall y \in H(x_1), y \in H(x_2)$. We then claim that there is a similar proof that if $x_2 * x_1' \in H$ then $\forall y \in H(x_2)$, $y \in H(x_1)$, which comes about by interchanging the roles of the symbols x_1 and x_2. The required result follows because if $p_1 \Rightarrow q_1$ is true and $p_2 \Rightarrow q_2$ is true then $p_1 \wedge p_2 \Rightarrow q_1 \wedge q_2$ is true. Though the statement $((p_1 \Rightarrow q_1) \wedge (p_2 \Rightarrow q_2)) \Rightarrow ((p_1 \wedge p_2) \Rightarrow (q_1 \wedge q_2))$ does not occur in the list of tautologies given in §2.4, it is easy to check that it is a tautology either by direct reference to its truth table or by construction from the tautologies listed in §2.4. The proof is presented with classical brevity.

Proof.   (i) It has previously been noted that $x_1 \in H(x_1)$. If $H(x_1) = H(x_2)$ then $x_1 \in H(x_2)$, i.e. $x_1 * x_2' \in H$. Similarly $x_2 * x_1' \in H$. Thus, by proof by parts, the condition is necessary.

(ii) If $x_1 * x_2' \in H$ and $y \in H(x_1)$, then $y * x_1' \in H$, and so

$$(y * x_1') * (x_1 * x_2') = y * x_2' \in H,$$

i.e.

$$y \in H(x_2).$$

Similarly, if $x_2 * x_1' \in H$ and $y \in H(x_2)$, then $y \in H(x_1)$. Thus the condition is sufficient.

Statements which follow from propositions by means of simple auxiliary arguments are often described as corollaries. We give a corollary to Proposition 4.5.

Corollary 4.6 Let $(H, *)$ be a subgroup of an abelian group $(G, *)$. If $H(x_1) = H(x_2)$ and $H(y_1) = H(y_2)$ then $H(x_1 * y_1) = H(x_2 * y_2)$.

Proof By Proposition 4.5, $x_1 * x_2' \in H$ and $y_1 * y_2' \in H$, so that

$$(x_1 * x_2') * (y_1 * y_2') \in H.$$

Since G is abelian,

$$(x_1 * x_2') * (y_1 * y_2') = (x_1 * y_1) * (x_2 * y_2)'.$$

Thus

$$(x_1 * y_1) * (x_2 * y_2)' \in H$$

and the result follows by the proposition.

The essence of this corollary is that $H(x * y)$ depends only on $H(x)$ and $H(y)$, not on the particular elements x and y. Another way to express the corollary is to say that the equation $H(x) \circ H(y) = H(x * y)$ defines a relation with domain $G/H \times G/H$, and range G/H. Moreover, the corollary ensures that this relation is a function, and so is a binary operation on G/H. We complete this section with a proof that $(G/H, \circ)$ is a group. It is called the *quotient group* of G by H.

Proposition 4.7. Let $(H, *)$ be a subgroup of an abelian group $(G, *)$. Let $H(x) = \{y \mid y * x' \in H\}$ and $G/H = \{H(x) \mid x \in G\}$. Define $H(x) \circ H(y) = H(x * y)$. Then $(G/H, \circ)$ is a group.

Proof (i) It follows from Corollary 4.6 that $\circ$ is a binary operation on G/H.

(ii) If $x, y, z \in G$ then

$$(H(x) \circ H(y)) \circ H(z) = H((x*y)*z)$$
$$= H(x*(y*z)) = H(x) \circ (H(y) \circ H(z)).$$

Hence the binary operation on G/H is associative.

(iii) If $x \in G$ then

$$H(x) \circ H(e) = H(x*e) = H(x)$$

and

$$H(e) \circ H(x) = H(e*x) = H(x).$$

Hence $H(e)$ is an identity element for the operation on G/H.

(iv) If $x \in G$ then

$$H(x') \circ H(x) = H(x'*x) = H(e)$$

and

$$H(x) \circ H(x') = H(x*x') = H(e).$$

Hence $H(x')$ is an inverse element for $H(x)$.

This completes the proof that $(G/H, \circ)$ is a group.

Exercises 4.3

1. Consider the binary operations f_1 to f_{14} introduced in §4.2. If X_n denotes the set on which f_n is acting, decide for each n whether (X_n, f_n) is a group.

2. Let $(G, *)$ be a group. Prove that if $(H_1, *)$ and $(H_2, *)$ are two subgroups of $(G, *)$, then $(H_1 \cap H_2, *)$ is a subgroup of $(G, *)$. Is $(H_1 \cup H_2, *)$ necessarily a subgroup of $(G, *)$?

3. Let $X = \{1, 2, 3, 4\}$ and let g be the binary operation of multiplication modulo 5, i.e. $g(x, y)$ is the remainder when $x \cdot y$ is divided by 5. Construct the block diagram for g and show that (X, g) is a group.

Investigate the corresponding situation when $X = \{1, 2, \ldots, n-1\}$ and when g is the operation of multiplication modulo n.

4. Let $X = \{0, 1, 2, \ldots, n-1\}$ and let f be the binary operation of additional modulo n, i.e. $f(x, y)$ is the remainder when $x+y$ is divided by n. Prove that (X, f) is a group. Construct the block

I

diagram for the case $n = 6$ and find as many subgroups of (X, f) as you can in this case.

5. Let $(G, *)$ and $(H, \circ)$ be two groups. Let $\square$ be a binary operation defined on $G \times H$ by

$$\forall g_1, g_2 \in G, \forall h_1, h_2 \in H, (g_1, h_1) \square (g_2, h_2) = (g_1 * g_2, h_1 \circ h_2).$$

Prove that $(G \times H, \square)$ is a group.

6. Let $(G, *)$ be a group for which G is a finite set. Let $(H, *)$ be a subgroup of $(G, *)$. Let $H(x_1)$ and $H(x_2)$ be two right cosets of $(H, *)$ in $(G, *)$. Consider the function f with domain $H(x_1)$ defined by $f: y \rightsquigarrow y * x_1' * x_2$, where y denotes an element of $H(x_1)$. Prove that the range of f is $H(x_2)$ and that f is a bijection from $H(x_1)$ to $H(x_2)$. Deduce that $H(x_1)$ and $H(x_2)$ have the same number of elements. Use the fact that the right cosets form a partition of G to prove that the number of elements in G is an integer multiple of the number of elements in H.

7. Let H denote the set of integer multiples of 3, and let $(Z, +)$ denote the group of integers under addition. Prove that $(H, +)$ is a subgroup of $(Z, +)$. Write down the right cosets of $(H, +)$ in $(Z, +)$ and verify, by constructing the block diagram, that $(Z/H, \circ)$ is a group, where $\circ$ denotes the operation described in Proposition 4.7.

4.4   METRIC SPACES

We turn now to the problem of developing a mathematical theory which can be applied to a wide collection of situations in which distance is measured. Although we shall begin by discussing 'distances' between physical objects, the resulting theory will be found to be valuable in studying 'distances' between mathematical objects such as functions.

The way that one measures the distance between two points depends on the use to which one will put the measurements. For some purposes one needs the distance between two points on the earth measured 'as the crow flies'. However, distances between towns are often measured along the shortest road routes for the convenience of motorists. Distances in a town with a one-way street system might be measured along permitted routes. Distances between two points on a circle might be measured along an arc of the circle or across a chord of the circle. In each of these cases, for a chosen set of points and a chosen unit of measurement one associates with each ordered pair of points of the set a positive real number.

In terms of the ideas of the previous chapter, we have a function ρ whose domain is some set of ordered pairs, and whose range is a set of real numbers. In practical problems involving metric properties one is concerned only with measuring distances between different objects. Thus it would seem that if X is the set of objects under consideration, the domain of ρ should be the set $X \times X \setminus \{(x, x) \mid x \in X\}$, and that the range of ρ should be a set of positive real numbers. However from the mathematical point of view this is a little complicated, and we can simplify matters by taking as the domain of ρ the set $X \times X$. We then have to decide what meaning to give to the symbol $\rho(x, x)$, and common sense dictates that if one should ever think of measuring the distance from x to itself, it ought to be zero. Thus the mathematical model consists so far of a set X and a function $\rho: X \times X \leadsto R$, with the properties that

$$\forall x \in X, \quad \rho(x, x) = 0 \quad \text{and} \quad \forall x, y \in X, \quad x \neq y \Rightarrow \rho(x, y) > 0.$$

The distinction here between decisions taken on practical grounds and those taken on mathematical grounds is important, and constantly appears in the process of constructing mathematical models.

For most ways of measuring commonly used the distance between two places does not depend on which of the two places one starts from. An exception to this occurs in certain distances measured in a town. The drive from A to B along one-way streets may be further than the drive back from B to A along other one-way streets. The more common situation can be expressed in terms of the function ρ by the statement, $\forall x, y \in X, \ \rho(x, y) = \rho(y, x)$. Similarly, the common idea that it is no worse to go direct between two points than via a third point can be expressed by the statement

$$\forall x, y, z \in X, \quad \rho(x, y) + \rho(y, z) \geqslant \rho(x, z).$$

In terms of the statements formulated above, most situations involving measurement of distance can be modelled in mathematics by means of the ideas of sets and functions. The restrictions placed on the function ρ certainly restrict the range of application of the model, but without these restrictions the task of calculating within the model would be too difficult. The model will be called a metric space.

Definition 4.8   A *metric space* is an ordered pair (X, ρ), where X is a set and ρ is a function from $X \times X$ to the real numbers such that

(i) $\forall x \in X, \quad \rho(x, x) = 0;$

(ii) $\forall x, y \in X, \quad x \neq y \Rightarrow \rho(x, y) > 0;$

(iii) $\forall x, y \in X, \quad \rho(x, y) = \rho(y, x);$

(iv) $\forall x, y, z \in X, \quad \rho(x, y) + \rho(y, z) \geqslant \rho(x, z).$

The function ρ is said to be a *metric* for the set X.

A metric space which is frequently applied to everyday problems is the coordinate plane R^2 with the Euclidean metric. If the notation (p_1, p_2) is used for a typical point of R^2, then the Euclidean metric is the function

$$\rho : R^2 \times R^2 \rightsquigarrow R,$$

$$\rho : ((p_1, p_2), (q_1, q_2)) \rightsquigarrow ((p_1 - q_1)^2 + (p_2 - q_2)^2)^{\frac{1}{2}}.$$

For every point (p_1, p_2),

$$\rho((p_1, p_2), (p_1, p_2)) = 0.$$

Moreover, if $(p_1, p_2) \neq (q_1, q_2)$, then

$$\rho((p_1, p_2), (q_1, q_2)) = \rho((q_1, q_2), (p_1, p_2)) > 0,$$

and so the first three properties for a metric space are satisfied. The fourth property now takes the special form that for every triple of points in R^2,

$$((p_1 - q_1)^2 + (p_2 - q_2)^2)^{\frac{1}{2}} + ((q_1 - r_1)^2 + (q_2 - r_2)^2)^{\frac{1}{2}}$$
$$\geqslant ((p_1 - r_1)^2 + (p_2 - r_2)^2)^{\frac{1}{2}}.$$

It is necessary to give an arithmetic proof of this inequality; it is not sufficient to appeal to the geometrical statement that the sum of the lengths of two sides of a triangle is greater than the length of the third side. The inequality can be proved directly using the fact that for any two positive real numbers, $u \geqslant v$ if and only if $u^{\frac{1}{2}} \geqslant v^{\frac{1}{2}}$. However, we shall prove a stronger inequality having other applications by making use of this fact together with the observation that a quadratic $ax^2 + bx + c$ with $a > 0$ is non-negative for all real values of x if and only if it has at most one real root, which occurs if and only if $b^2 - 4ac \leqslant 0$.

Proposition 4.8 *Cauchy's Inequality* Given any natural number n, and any collection of $2n$ real numbers $\{u_1,..,u_n, v_1,\ldots,v_n\}$,

$$\left(\sum_{i=1}^{n} u_i v_i\right)^2 \leqslant \left(\sum_{i=1}^{n} u_i^2\right)\left(\sum_{i=1}^{n} v_i^2\right).$$

Proof Since

$$x^2 \sum_{i=1}^{n} u_i^2 + 2x \sum_{i=1}^{n} u_i v_i + \sum_{i=1}^{n} v_i^2 = \sum_{i=1}^{n} (u_i x + v_i)^2 \geqslant 0,$$

for all real values of x, it follows that

$$\left(2\sum_{i=1}^{n} u_i v_i\right)^2 - 4\left(\sum_{i=1}^{n} u_i^2\right)\left(\sum_{i=1}^{n} v_i^2\right) \leqslant 0,$$

from which the required result may be obtained.

Proposition 4.9 *Minkowski's Inequality* Given any natural number n, and any collection of $2n$ real numbers $\{u_1,\ldots,u_n, v_1,\ldots,v_n\}$

$$\left(\sum_{i=1}^{n} (u_i + v_i)^2\right)^{\frac{1}{2}} \leqslant \left(\sum_{i=1}^{n} u_i^2\right)^{\frac{1}{2}} + \left(\sum_{i=1}^{n} v_i^2\right)^{\frac{1}{2}}.$$

Proof It follows from Cauchy's inequality, and from the property of squares roots noted above, that

$$\left|\sum_{i=1}^{n} u_i v_i\right| \leqslant \left(\sum_{i=1}^{n} u_i^2\right)^{\frac{1}{2}} \left(\sum_{i=1}^{n} v_i^2\right)^{\frac{1}{2}}.$$

Thus

$$\sum_{i=1}^{n} (u_i + v_i)^2 = \sum_{i=1}^{n} u_i^2 + 2\sum_{i=1}^{n} u_i v_i + \sum_{i=1}^{n} v_i^2$$

$$\leqslant \sum_{i=1}^{n} u_i^2 + 2\left|\sum_{i=1}^{n} u_i v_i\right| + \sum_{i=1}^{n} v_i^2$$

$$\leqslant \sum_{i=1}^{n} u_i^2 + 2\left(\sum_{i=1}^{n} u_i^2\right)^{\frac{1}{2}} \left(\sum_{i=1}^{n} v_i^2\right)^{\frac{1}{2}} + \sum_{i=1}^{n} v_i^2$$

$$= \left[\left(\sum_{i=1}^{n} u_i^2\right)^{\frac{1}{2}} + \left(\sum_{i=1}^{n} v_i^2\right)^{\frac{1}{2}}\right]^2$$

and the required result follows on taking square roots.

Now, putting $n = 2$, $u_i = p_i - q_i$ and $v_i = q_i - r_i$ in Minkowski's inequality, the inequality required for the Euclidean metric in R^2 is obtained. Similarly, if $(p_1, p_2, \ldots, p_n)$ is a typical point of the cartesian product R^n of R with itself n times, then the Minkowski inequality also shows that the function

$$\rho: R^n \times R^n \rightsquigarrow R, \; \rho: ((p_1, p_2, \ldots, p_n), (q_1, q_2, \ldots, q_n)) \rightsquigarrow \left[\sum_{i=1}^{n} (p_i - q_i)^2 \right]^{\frac{1}{2}}$$

satisfies condition (iv) for a metric. The first three conditions are also satisfied by this function, and so it is a metric for R^n. The metric space (R^n, ρ) is called n-dimensional Euclidean space.

The coordinate plane R^2 admits other metrics in addition to the Euclidean metric. For example, the function

$$\rho: R^2 \times R^2 \rightsquigarrow R, \; \rho: ((p_1, p_2), (q_1, q_2)) \rightsquigarrow |p_1 - q_1| + |p_2 - q_2|$$

clearly satisfies the first three conditions for a metric. Since for any two real numbers u and v, $|u| + |v| \geqslant |u + v|$, given three points in R^2 we have

$$(|p_1 - q_1| + |p_2 - q_2|) + (|q_1 - r_1| + |q_2 - r_2|) \geqslant (|p_1 - r_1| + |p_2 - r_2|).$$

Thus the function satisfies the fourth condition, and is therefore a metric for R^2. The restriction of this metric to a suitable finite subset of R^2 is an appropriate model for application to problems involving transport in a town which is laid out with two sets of parallel roads intersecting at right angles.

In many measuring situations one dismisses distances which are too large and concentrates on the measurement of distances which are within a reasonable range for the job in hand. This has an analogue in the theory of metric spaces. The next proposition shows that if in a metric space all distances greater than a fixed number k are replaced by that distance, then one still has a metric space.

Proposition 4.10 Let (X, ρ) be a metric space and let k be a positive real number. Then the function ρ^* defined by

$$\rho^*: X \times X \rightsquigarrow R,$$

$$\rho^*: (x, y) \rightsquigarrow \min (\rho(x, y), k),$$

is a metric for X.

Proof (i) $\forall x \in X$, $\rho^*(x, x) = \rho(x, x) = 0$.

(ii) $\forall x, y \in X$, $x \ne y \Rightarrow \rho(x, y) > 0$. Since also $k > 0$,

$$\min(\rho(x, y), k) = \rho^*(x, y) > 0.$$

(iii) $\forall x, y \in X$, $\min(\rho(x, y), k) = \min(\rho(y, x), k)$.
Hence $\rho^*(x, y) = \rho^*(y, x)$.

(iv) $\forall x, y, z \in X$,

$$\begin{aligned}
\rho^*(x, y) + \rho^*(y, z) &= \min(\rho(x, y), k) + \min(\rho(y, z), k) \\
&\geqslant \min(\rho(x, y) + \rho(y, z), \rho(x, y) + k, \rho(y, z) + k, k + k) \\
&\geqslant \min(\rho(x, z), \rho(x, y) + k, \rho(y, z) + k, k + k) \\
&\geqslant \min(\rho(x, z), k) = \rho^*(x, z).
\end{aligned}$$

This completes the proof that ρ^* is a metric for X.

As an example of the application of metric spaces to other branches of mathematics, we consider certain metric spaces of functions. Let $[a, b]$ denote the interval $\{x \mid x \in R \text{ and } a \leqslant x \leqslant b\}$, and let $\mathscr{F}$ denote the set of all polynomial functions with domain $[a, b]$, i.e. the set of all functions of the form

$$f: [a, b] \rightsquigarrow R,$$

$$f: x \rightsquigarrow a_n x^n + \ldots + a_1 x + a_0.$$

We shall make use of the fact that each such function f can be integrated and differentiated, and that $|f(x)|$ has a maximum value in $[a, b]$. The set $\mathscr{F}$ admits many metrics of which we introduce two here.

Proposition 4.11 The functions

$$\rho_1: \mathscr{F} \times \mathscr{F} \rightsquigarrow R, \quad \rho_1: (f, g) \rightsquigarrow \max_{a \leqslant x \leqslant b} |f(x) - g(x)|,$$

$$\rho_2: \mathscr{F} \times \mathscr{F} \rightsquigarrow R, \quad \rho_2: (f, g) \rightsquigarrow \int_a^b |f(x) - g(x)| \, dx,$$

are metrics for the set $\mathscr{F}$ of polynomial functions with domain $[a, b]$.

Proof (i) If $f = g$ then $\forall x \in [a, b]$, $f(x) = g(x)$, so that

$$\rho_1(f, g) = \rho_2(f, g) = 0.$$

(ii) If $f \neq g$ then $\exists x_0 \in [a,b]$ such than $f(x_0) \neq g(x_0)$, whence $|f(x_0)-g(x_0)| > 0$. Thus

$$\rho_1(f,g) \geqslant |f(x_0)-g(x_0)| > 0.$$

Moreover, since the function $f-g$ is not identically zero, it is a polynomial of some specific degree n, and so has no more than n zeros in $[a,b]$. Thus

$$\rho_2(f,g) = \int_a^b |f(x)-g(x)| \, dx > 0.$$

(iii) Given two polynomial functions f and g,

$$\forall x \in [a,b], |f(x)-g(x)| = |g(x)-f(x)|.$$

Hence $\rho_1(f,g) = \rho_1(g,f)$ and $\rho_2(f,g) = \rho_2(g,f)$.

(iv) Let f,g,h be three polynomial functions with domain $[a,b]$. Then

$$\forall x \in [a,b], |f(x)-h(x)| = |f(x)-g(x)+g(x)-h(x)|$$
$$\leqslant |f(x)-g(x)|+|g(x)-h(x)|.$$

Thus

$$\max_{a \leqslant x \leqslant b} |f(x)-h(x)| \leqslant \max_{a \leqslant x \leqslant b} (|f(x)-g(x)|+|g(x)-h(x)|)$$
$$\leqslant \max_{a \leqslant x \leqslant b} |f(x)-g(x)| + \max_{a \leqslant x \leqslant b} |g(x)-h(x)|,$$

i.e. $\rho_1(f,h) \leqslant \rho_1(f,g)+\rho_1(g,h)$.

It follows also that

$$\int_a^b |f(x)-h(x)| \, dx \leqslant \int_a^b (|f(x)-g(x)|+|g(x)-h(x)|) \, dx$$
$$= \int_a^b |f(x)-g(x)| \, dx + \int_a^b |g(x)-h(x)| \, dx,$$

i.e. $\qquad \rho_2(f,h) \leqslant \rho_2(f,g)+\rho_2(g,h)$.

This completes the proof that ρ_1 and ρ_2 are metrics for $\mathscr{F}$.

These metric spaces of polynomial functions, and other metric spaces of functions, are of great value in advanced calculus, particularly in the study of the approximation of one type of function by another type of function, where clearly an idea of distance between functions can provide a measure of closeness. We shall use these ideas here in discussing a particular aspect of the use of calculus in science.

When some physical variable such as the distance travelled by an object is measured to obtain a relationship between distance travelled and time, only a finite number of measurements is taken, and there is a certain degree of error in each of the measurements. One commonly fits a polynomial function to the readings taken, and then uses differential calculus to find the rate of change of the variable in time, which in the above example would be the velocity of the object. Let us see what effect the errors in the readings have on this process. Suppose that f and g are two polynomial functions which are equally good fits for the readings. It is reasonable to interpret this intuitive remark by the mathematical statement that

$$\max_{a \leqslant x \leqslant b} |f(x) - g(x)| < \varepsilon,$$

i.e.

$$\rho_1(f, g) < \varepsilon,$$

where ε is some positive number determined by the possible errors in the readings. Now what does this tell us about the variation between the derivatives f' and g' of f and g? Unfortunately, nothing, for it is possible for $\rho_1(f', g')$ to be very large even when $\rho_1(f, g)$ is small. This is indicated in Fig. 4.1.

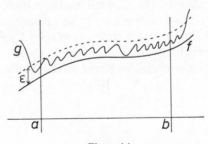

Figure 4.1

Of course, in the diagram, the degree of the polynomial function g is very much larger than the degree of the polynomial function f, and their values differ greatly outside the interval $[a, b]$. In that interval however the values of f and g are within ε of each other, whereas the values of f' and g' differ greatly. When the differential

K

calculus is used in science it is under the rather strong assumption that if a physical variable is represented approximately by the function $f(t)$, then the rate of change of the variable is represented approximately by the derived function $f'(t)$. This is a restrictive condition placed on the use of differential calculus in science; it is not a consequence of a mathematical theorem in the differential calculus.

For the use of the integral calculus there are no such troubles. One of the properties of integrals is that for a positive function f,

$$\int_a^b f(x)\,dx \leqslant (b-a)\max_{a\leqslant x\leqslant b} f(x).$$

Thus for any two polynomial functions f and g,

$$\int_a^b |f(x)-g(x)|\,dx \leqslant (b-a)\max_{a\leqslant x\leqslant b}|f(x)-g(x)|,$$

i.e.

$$\rho_2(f,g) \leqslant (b-a)\rho_1(f,g).$$

It follows that if $\rho_1(f,g) < \varepsilon$, then

$$\rho_2(f,g) < (b-a)\,\varepsilon.$$

As a consequence of this result, from an approximate knowledge of the rate of change of a physical variable one can calculate the approximate values of the physical variable itself. For example, given the approximate acceleration of an object during an interval of time one can calculate its approximate velocity during that interval. In this case the result follows from a theorem about integral calculus and is not a restrictive condition placed on the use of integral calculus in science.

Exercises 4.4

1. Discuss the extent to which a metric space might be considered a satisfactory model for airline times between airports.

2.(a) Show that the functions ρ_1, ρ_2 defined below are both metrics for the coordinate plane R^2.

$$\rho_1 : ((x_1,y_1),(x_2,y_2)) \rightsquigarrow |x_1-x_2|+|y_1-y_2|,$$

$$\rho_2 : ((x_1,y_1),(x_2,y_2)) \rightsquigarrow \max(|x_1-x_2|,|y_1-y_2|).$$

Let E_ρ denote the unit circle in (R^2, ρ), i.e.,

$$E_\rho = \{(x, y) \mid \rho((x, y), (0, 0)) = 1\}.$$

Draw diagrams of E_{ρ_1} and E_{ρ_2}.

(b) In (R^2, ρ) we say that the point $A = (x, y)$ is ρ-between the points $A_1 = (x_1, y_1)$ and $A_2 = (x_2, y_2)$ if and only if

$$\rho(A_1, A_2) = \rho(A_1, A) + \rho(A, A_2).$$

Draw diagrams of the two sets

$$\{A \mid A \text{ is } \rho_1\text{-between } A_1 \text{ and } A_2\},$$
$$\{A \mid A \text{ is } \rho_2\text{-between } A_1 \text{ and } A_2\}.$$

(c) In (R^2, ρ) we call the set

$$P_\rho = \{A \mid \rho(A, A_1) = \rho(A, A_2)\}$$

the ρ-perpendicular bisector of $\{A_1, A_2\}$. Draw diagrams of the sets P_{ρ_1} and P_{ρ_2}.

3. Let (X, ρ) be a metric space. A point y will be said to lie between points x and z in X if the three points are distinct and if $\rho(x, z) = \rho(x, y) + \rho(y, z)$. The notation xyz will be used to denote that y lies between x and z. Decide on the truth of the following propositions.

(i) If pqr and prs then pqs.
(ii) If pqr and prs then qrs.
(iii) If pqr and qrs then pqs.

4. In Cauchy's inequality and in Minkowski's inequality investigate the circumstances under which equality occurs.

5. Let (X, ρ) be a metric space. Define a function $\sigma: X \times X \rightsquigarrow R$ by

$$\sigma(x, y) = \frac{\rho(x, y)}{1 + \rho(x, y)}.$$

Show that σ is a metric for X and that $\forall x, y \in X, \sigma(x, y) < 1$.

6. Let (X, ρ_1) and (X, ρ_2) be metric spaces. Let $\rho: X \times X \rightsquigarrow R$ be defined by

$$\rho(x, y) = \rho_1(x, y) + \rho_2(x, y).$$

Show that (X, ρ) is a metric space.

7. Let (X_1, ρ_1) and (X_2, ρ_2) be metric spaces. Let $\rho: (X_1 \times X_2) \times (X_1 \times X_2) \rightsquigarrow R$ be defined by

$$\rho((x_1, x_2), (y_1, y_2)) = [\rho_1(x_1, y_1)^2 + \rho_2(x_2, y_2)^2]^{\frac{1}{2}}.$$

Show that $(X_1 \times X_2, \rho)$ is a metric space.

8. Let $f(x) = 2x - 2x^2 + 3x^3$, and $g(x) = x + x^2 - x^3$. Show that for all values of x in the interval $[-1/120, 1/120]$, $|f(x) - g(x)| \leqslant 1/100$. but that $|f'(x) - g'(x)| \geqslant 19/20$.

4.5 MEASURE SPACES

We shall be concerned in this section with a mathematical model associated with certain kinds of measurement. Consider the examples of measurement of lengths in terms of centimetres, measurement of volumes in terms of litres, measurement of quantities of liquids in terms of litres, and measurement of mass in grams. In ordinary situations, given two separate objects whose measurements in a given system are x and y units respectively, the object formed by combining them has the measurement $(x + y)$ units. One can only indicate what is ordinary in comparison with extraordinary measurement situations. The sum of the volume of a brick and the volume of a block of rubber will be larger than the volume of the combined object formed by putting the brick on the rubber block: the sum of the length of a piece of elastic and the length of a metal rule will be less than the length of the compound object formed by hanging the metal rule on the end of the piece of elastic. These you will regard as exceptional situations to which the usual rule of addition of measurements does not apply. Nevertheless, it does highlight the fact that in a very large number of situations, one does consider measurements to be additive.

Another measurement often considered is that of temperature. This differs in many respects from the situations described above. Firstly the scale of measurement may include negative numbers, and more important, difficulties arise when we wish to combine objects. For example, if a heated brass weight is placed in a beaker of cold water, it does not make sense immediately to talk of the temperature of the combined object (brass weight in beaker of water) so formed. One must wait until thermal equilibrium is reached and then measure the temperature. Neither is temperature additive in this situation.

This is overcome in physical problems by considering quantities of heat, which permit the concept of additivity to be used. The two types of measurement then have to be related, and in this case it is done through the auxiliary concept of thermal capacity. This last example emphasizes the use which is made of additivity, even in situations where at first sight it does not apply, and points to the fact that in additive situations non-negative scales of measurement are used. This, together with the large number of situations which are additive, makes it worthwhile to use the notion in constructing a mathematical model to describe these processes of measurement.

In addition to combining objects, one also considers measurement situations where subtraction is involved. An example would be the calculation of areas in a plot of land to be used for various purposes, where subtraction would be used to find the residue when a number of portions had been allocated. Clearly there will be many situations of this kind involving differences in length, mass or volume.

To summarize, in any system of measurement, be it of length, of volume or of mass, one conceives of a collection of objects to each of which is ascribed a measure expressed in chosen units. The combination of two objects is ascribed a measure and so is the remainder when part of an object is removed. In set theoretic terms, one must be able to consider with any pair of objects the measure of their union and their difference. In order to obtain a model which can be applied to any of the common systems of measurement described above we shall drop any reference to units of measurement and just use the non-negative real numbers. Thus we are led to consider the following mathematical system.

Definition 4.9 A *measure space* is an ordered triple $(U, \mathcal{A}, m)$ such that

   (i) U is a non-empty set,

  (ii) $\mathcal{A}$ is a non-empty subset of $\mathcal{P}(U)$,

 (iii) $\forall A, B \in \mathcal{A}, A \cup B \in \mathcal{A}$ and $A \backslash B \in \mathcal{A}$,

 (iv) $m: \mathcal{A} \rightsquigarrow R$ is a real valued function,

  (v) $\forall A \in \mathcal{A}, m(A) \geqslant 0$,

 (vi) $\forall A, B \in \mathcal{A}$, if $A \cap B = \varnothing$ then $m(A \cup B) = m(A) + m(B)$.

The concept of a measure space can be applied to any situation in which one believes that a collection of objects can be measured additively in chosen units. One associates with each member A of $\mathscr{A}$ an object to be measured and one says that the measure of the object is $m(A)$ units.

Some elementary consequences of the definition are proved below. None of them is remarkable when considered in relation to examples of measurements of length, volume or mass. They serve to show that the model reflects familiar properties other than those stated explicitly in Definition 4.9. We give only the discursive forms of the proofs.

Proposition 4.12 $\varnothing \in \mathscr{A}$ and $m(\varnothing) = 0$.

Proof By (ii), $\exists A \in \mathscr{A}$; and by (iii), $A \backslash A = \varnothing \in \mathscr{A}$. Since $A \cap \varnothing = \varnothing$, it follows from (vi) that $m(A \cap \varnothing) = m(A) + m(\varnothing)$. Moreover, since $A \cap \varnothing = A$, $m(A \cap \varnothing) = m(A)$. Thus $m(A) + m(\varnothing) = m(A)$, i.e. $m(\varnothing) = 0$.

Proposition 4.13 If $A, B \in \mathscr{A}$, then $m(A \backslash B) = m(A) - m(A \cap B)$.

Proof By exercise 1 of §2.7, $A = (A \backslash B) \cup (A \cap B)$ and $(A \backslash B) \cap (A \cap B) = \varnothing$. Also, you are asked to show in exercise 2 below that $A \cap B \in \mathscr{A}$. Then by (vi), $m(A) = m(A \backslash B) + m(A \cap B)$, i.e. $m(A \backslash B) = m(A) - m(A \cap B)$.

Corollary 4.14 If $B \subseteq A$ then $m(A \backslash B) = m(A) - m(B)$.

Proof Since by Proposition 2.3, $B \subseteq A$ implies $A \cap B = B$, $B \subseteq A$ also implies $m(A \backslash B) = m(A) - m(A \cap B) = m(A) - m(B)$.

Corollary 4.15 If $U \in \mathscr{A}$ and $A \in \mathscr{A}$, $m(A^c) = m(U) - m(A)$.

Proof Replace A and B in Proposition 4.13 by U, A respectively.

Proposition 4.16 If $B \subseteq A$ then $m(B) \leqslant m(A)$.

Proof By Corollary 4.14, if $B \subseteq A$ then $m(A) - m(B) = m(A \backslash B)$. Also by (v), $m(A \backslash B) \geqslant 0$, and hence $m(B) \leqslant m(A)$.

We now give two extensions of property (vi). The first concerns the measure of the union of two sets which need not be disjoint. The second concerns the measure of the union of more than the disjoint sets.

Proposition 4.17 If $A, B \in \mathcal{A}$ then $m(A \cup B) = m(A) + m(B) - m(A \cap B)$.

Proof By exercise 1 of §2.7, $A \cup B = B \cup (A \backslash B)$ and $B \cap (A \backslash B) = \varnothing$, and so by (vi), $m(A \cup B) = m(B) + m(A \backslash B)$. Thus by Proposition 4.13, $m(A \cup B) = m(B) + m(A) - m(A \cap B)$.

Two sets A and B have been called disjoint if $A \cap B = \varnothing$. A collection of sets $\{A_1, A_2, \ldots, A_n\}$ is said to be a *pairwise disjoint* collection if $i \neq j$ implies $A_i \cap A_j = \varnothing$. Some authors describe this condition by saying that the sets are *mutually exclusive*.

Proposition 4.18 For all natural numbers $n \geqslant 2$, and all collections of sets $\{A_1, A_2, \ldots, A_n\}$, if $A_1, A_2, \ldots, A_n$ are pairwise disjoint then

$$m(A_1 \cup \ldots \cup A_n) = m(A_1) + m(A_2) + \ldots + m(A_n).$$

Proof The proposition is of the form $\forall n \in N \backslash \{1\}$, $P(n)$. Let $W = \{n \mid P(n) \text{ is true}\}$. Now the statement $P(2)$ is: if $A_1 \cap A_2 = \varnothing$ then $m(A_1 \cup A_2) = m(A_1) + m(A_2)$. This is true by condition (vi), hence $2 \in W$. Now suppose that $n \in W$. Suppose also that the sets $A_1, A_2, \ldots, A_n, A_{n+1}$ are pairwise disjoint. Then by exercise 6 of §3.8,

$$(A_1 \cup \ldots \cup A_n) \cap A_{n+1} = (A_1 \cap A_{n+1}) \cup \ldots \cup (A_n \cap A_{n+1})$$
$$= \varnothing \cup \ldots \cup \varnothing = \varnothing.$$

Hence by (vi),

$$m((A_1 \cup \ldots \cup A_n) \cup A_{n+1}) = m(A_1 \cup \ldots \cup A_n) + m(A_{n+1}).$$

Since $n \in W$, $m(A_1 \cup \ldots \cup A_n) = m(A_1) + \ldots + m(A_n)$. Thus $n+1 \in W$. Now by Peano axiom (iv), since $2 \in W$ and $\forall n, n \in W \Rightarrow n+1 \in W$, it follows that $W = N \backslash \{1\}$. This completes the proof.

Exercises 4.5

1. Discuss the extent to which a measure space might be considered a satisfactory model for sweetness of batches of orange juice.

2. Let U be a non-empty set and suppose that $\mathscr{A}$ is a non-empty subset of $\mathscr{P}(U)$ satisfying (iii) above, i.e. $\forall A, B \in \mathscr{A}$, $A \cup B \in \mathscr{A}$ and $A \backslash B \in \mathscr{A}$. Show that

   (a) $\forall A, B \in \mathscr{A}$, $A \cap B \in \mathscr{A}$,

   (b) $\forall A, B \in \mathscr{A}$, $A \triangle B \in \mathscr{A}$,

   (c) If $A_1 \in \mathscr{A}, A_2 \in \mathscr{A}, \ldots, A_n \in \mathscr{A}$ then $\bigcup\limits_{i=1}^{n} A_i \in \mathscr{A}$ and $\bigcap\limits_{i=1}^{n} A_i \in \mathscr{A}$.

3. Let U be a non-empty set and suppose that $\mathscr{A}$ and $\mathscr{B}$ are both subsets of $\mathscr{P}(U)$ satisfying (iii) above. Does $\mathscr{A} \cap \mathscr{B}$ satisfy (iii)? Do $\mathscr{A} \cup \mathscr{B}$ and $\mathscr{A} \backslash \mathscr{B}$ satisfy (iii)? Does $\mathscr{P}(U)$ itself satisfy (iii)?

4. Show that in any measure space,

$$m(A \cup B \cup C) = m(A) + m(B) + m(C) - m(A \cap B)$$
$$- m(A \cap C) - m(B \cap C) + m(A \cap B \cap C).$$

Can you generalize this rule?

5. Suppose U is a finite set, and $\mathscr{A} = \mathscr{P}(U)$. If $A \in \mathscr{A}$ let $m(A)$ denote the number of elements in A. Show that $(U, \mathscr{A}, m)$ is a measure space.

6. Analysis of a recent questionnaire showed that of a group of mathematics students, 80 liked pure mathematics, 90 liked applied mathematics, and 60 liked statistics; 70 liked both pure and applied, 40 liked pure and statistics, whilst 50 liked applied and statistics; 30 students liked all three subjects. How many students completed the questionnaire?

5
PROBABILITY THEORY

5.1 INTRODUCTION

The two early fields of study for statistics and probability were games of chance and economic and social statistics. In 1663, the Italian mathematician and gambler Gerolamo Cardano published the results of his calculation of the odds involved in games with dice. At about the same time the French mathematicians Blaise Pascal and Pierre Fermat were corresponding with each other about questions raised by another gambler, while in England mathematical studies of insurance and economic statistics were being pursued. Techniques of statistics were subsequently developed and the fields of application extended to include sampling for errors, correlation studies, genetics, astronomic and sub-atomic physics, educational testing, business planning, agricultural experiments, public opinion polls, and many others.

The special techniques and terminologies used in these applications depend on a knowledge of the physical circumstances involved as well as of various kinds of mathematical techniques. For each kind of application a detailed mathematical model is developed, embodying the concepts common to all statistical situations and also special auxiliary concepts appropriate to the particular application.

It is the object of this chapter to show how the set theoretic notions of the previous chapters can be used in the construction of a basic mathematical model of statistical situations, and to show how problems in statistics have led to the introduction of new terminologies and of auxiliary concepts.

In applications for which the appropriate auxiliary concepts are expressed in algebraic or analytic terms one loses sight of the basic set theoretic model and concentrates on algebraic or analytic techniques. Nevertheless, the problem of deciding which algebraic expressions or which functions are appropriate detailed models for particular applications bears a close resemblance to the problem of deciding on a description of the basic set theoretic model.

All the illustrations will involve games and experiments for which a sequence of trials can be envisaged, and probabilities of events will be closely related to their frequencies of occurrence in a long sequence of trials. In everyday language we speak also of probabilities of non-repeatable events. The statements are not expressed in numerical terms, but in ways such as 'this will probably happen' and 'that will probably not happen'. The basic mathematical model can be applied to some such situations as well as to repeatable games, although we shall not discuss here the relationship between the model and non-repeatable events.

5.2 FINITE PROBABILITY SPACES

The types of games which initiated the study of probabilities involved dealing cards from a pack, casting a die, casting several dice together, tossing a coin or coins and similar activities. We shall attempt to develop a model of these games expressed in set theoretic language. After a run of any one of these games one notes only a few of the important characteristics of the result. For example, after dealing a card one notes the number and suit of the card, but not its precise position on the table. After tossing a coin one notes which side of the coin is uppermost, using the familiar nomenclature of heads and tails; but one does not note its precise position on the table. On the other hand, in the game of push-penny one notes whether a coin comes to rest in one of ten regions, and one does not note which side of the coin is uppermost.

When a die is cast one normally notes the number of spots on the face which is uppermost when the die comes to rest. But a cast die may roll off the table or come to rest in a hole so that there is no face which is clearly on top. It is possible, though less likely, that a tossed coin might come to rest on its edge, or be caught in mid-air by a bird and carried away. There are two ways in which one can deal with such hazards. Firstly, one can consider only trials or runs of a game which have a regular outcome. Secondly, one can augment the set of results of a game to be considered by the inclusion of one described as 'mis-run' or 'foul' or possibly by the inclusion of a number of results which take into account a variety of possible hazards. In either case one decides on a collection of acceptable results of a game such that for each acceptable run of the game precisely one of the results occurs.

The words or symbols which are noted after a run of a game will be called the *outcomes* or the *elementary events* of the game. In the rest of this section, except where explicitly mentioned, the possible outcomes of casting a die will be 1, 2, 3, 4, 5, 6 and the possible outcomes of tossing a coin will be *h, t*. A player who tries to cast a die for an even number will be successful if the outcome of the trial is either 2 or 4 or 6. Again a player who tries to toss two coins together so that both show the same face will be successful if the outcome of the trial is *h* and *h*, or *t* and *t*. Such results will be called *compound events* or *events*.

The games described above all have the property that the elementary events form a finite set of which the events are subsets. The set of elementary events will be called the *sample space* for the game. In other statistical situations the set of acceptable outcomes of a trial may be infinite or be treated as infinite. For example, one can estimate the average height of an undergraduate by averaging the results of a sufficiently long sequence of repetitions of the action 'pick an undergraduate at random and record his height'. The heights are likely to be recorded to the nearest half inch (or centimetre), and they are likely to lie between 4 feet and 8 feet (1·25 m and 2·5 m), so there will be a finite number of outcomes. For some purposes it will be better to record the heights to the nearest inch to reduce the possible number of outcomes. For other purposes it will be better to act as though every real number is a possible outcome and to use calculus to analyse the results. The description of mathematical models for statistical situations in which the real numbers are taken as the possible outcomes is a natural development of the description of models for situations in which only a finite number of elementary events is chosen. We concentrate here on the latter.

Each of the games described at the beginning of this section has the second property that one can conceive of a sequence of independent runs of the game, and that for any pair of sufficiently long sequences of runs the proportion of times that any particular event will occur will be about the same. For example, in the absence of any evidence that the coins and dice involved have been loaded, one would expect that in any sequence of tosses of a coin about half the outcomes would be heads and in any sequence of throws of a die about one sixth of the outcomes would be sixes. The relative frequencies of compound events can be calculated from the relative frequencies of the elementary events. Thus in a sequence of throws of a die about half the outcomes would be even numbers, since the numbers 2, 4 and 6 would each occur about one sixth of the times.

If one suspected that a coin had been loaded then one would expect that in any sequence of tosses of the coin the proportion of heads would approximate to some number $p(h)$ while the proportion of tails would approximate to some number $p(t)$ such that $p(h) + p(t) = 1$. The actual numbers $p(h)$ and $p(t)$ could be taken as a measure of loading of the coin. Similarly, if one suspected that a die had been loaded then one would expect that in any sequence of casts of the die the proportions of occurrences of the elementary events would approximate to six numbers such that $p(1) + p(2) + p(3) + p(4) + p(5) + p(6) = 1$. Examination of the proportion of occurrences of a compound event in any one sequence of throws of the die indicates that it will approximate to the sum of the numbers $p(i)$ taken over those elementary events i which make up the compound event. For example the proportion of occurrences of even numbers will approximate to $p(2) + p(4) + p(6)$.

Whether the coin and die are thought to be fair or loaded, the outcomes of any run of a game with them will be uncertain, but will be subject to the restrictions imposed by our belief in the regularity of behaviour over a sequence of runs. Many other games and experiments have the same kind of behaviour. They are called *random games* and *random experiments*.

Satisfactory models for random games which have finitely many outcomes need to embody the two characteristics described above. Thus we consider as models, pairs (S, p) where S is a finite set and

$p : S \rightsquigarrow R$ is a function such that $\forall x \in S, p(x) \geqslant 0$ and $\sum_{x \in S} p(x) = 1$.

When used in this context the set theoretic symbols S, $E \subseteq S$, $E \cup F$, etc., are often described in a terminology peculiar to probability theory and statistics, which is controlled by the interpretation put on the symbol for a variable. The following list gives some of the more common symbols with their descriptions in general set theory and in probability.

The whole set S and the null set $\varnothing$ are subsets of S, and so are to be regarded as events. Since the set S consists of all possible out-

Symbols	Set theory description	Probability description
S	Universe of discourse	Sample space
$\mathscr{P}(S)$	Power set of S	Set of events
$x \in S$	x is an arbitrary element of S	x is the outcome of an arbitrary run of the game
$E \in \mathscr{P}(S)$	E is a subset of S	E is an event
$x \in E$	x is an element of E	Event E occurs
E^c	Complement of E in S	Event 'not E'
$E \cup F$	Union of E and F	Event E or F
$E \cap F$	Intersection of E and F	Event E and F
$E \subseteq F$	E is a subset of F	E implies F

comes of a game, the event S must occur for every acceptable trial of the game. Similarly, the event $\varnothing$ never occurs. In the language of probability, S is called the *certain event* and $\varnothing$ is called the *impossible event*.

For each elementary event $x \in S$, $p(x)$ is called the *probability of x*. Following the discussion of proportions of occurrences of compound events in a sequence of runs of a game, it is reasonable to extend the model for a random game by the introduction of a function $P: \mathscr{P}(S) \rightsquigarrow R$ defined in terms of the function p by summation. Precisely,

$$\forall E \in \mathscr{P}(S), P(E) = \sum_{x \in E} p(x).$$

Then $P(E)$ is called the *probability of E*. It follows from the conditions on p that $\forall E \in \mathscr{P}(S), P(E) \geqslant 0$, and that $P(S) = 1$. Moreover, $E \cap F = \varnothing$ implies $P(E \cup F) = P(E) + P(F)$. To prove this latter property, the elements of E and F are numbered so that $E = \{s_1, \ldots, s_n\}$ and $F = \{s_{n+1}, \ldots, s_m\}$, where $i \neq j$ implies $s_i \neq s_j$. It then follows that

$$P(E \cup F) = \sum_{i=1}^{m} p(s_i) = \sum_{i=1}^{n} p(s_i) + \sum_{i=n+1}^{m} p(s_i) = P(E) + P(F).$$

Thus the triple $(S, \mathscr{P}(S), P)$ satisfies the conditions for a measure space as defined in §4.5 and the additional condition $P(S) = 1$. It follows that the results of the propositions of that section apply immediately to the present situation.

Given any measure space of the form $(S, \mathscr{P}(S), P)$ such that

$P(S) = 1$ we can define a function $p: S \rightsquigarrow R$ by the rule $\forall x \in S$, $p(x) = P(\{x\})$. This function has the two properties $\forall x \in S, p(x) \geqslant 0$ and $\sum_{x \in S} p(x) = 1$. Thus if S is finite, it does not matter which of the two functions $p: S \rightsquigarrow R$ and $P: \mathscr{P}(S) \rightsquigarrow R$ is specified first and the other defined from it. However, if S is infinite, while p can be defined satisfactorily in terms of P, the attempt to define P from p leads to the requirement that an infinite sum of zeros should yield a positive number and so has to be abandoned. We are guided by the prospect of future extensions of the theory to the case of infinite sample spaces to choose the following definition of a finite probability space as the model for random games and experiments with finite sample spaces.

Definition 5.1   *A finite probability space* is an ordered pair (S, P) such that

   (i) S is a finite set,
  (ii) $P: \mathscr{P}(S) \rightsquigarrow R$ is a function,
 (iii) $P(S) = 1$,
  (iv) $\forall A \subseteq S, P(A) \geqslant 0$.
   (v) $\forall A, B \subseteq S$, if $A \cap B = \varnothing$ then $P(A \cup B) = P(A) + P(B)$.

The condition (iii) is sometimes called the *normalization condition*, and a finite probability space is sometimes described as a finite normalized measure space.

We are concerned not only with the investigation of the logical consequences of this definition, but also with the problem of deciding for which situations finite probability spaces form satisfactory models and with the problem of choosing specific spaces S and specific functions P appropriate to specific applications. Experience is the main guide, but the experience of others can be boiled down to five rules which a novice may take as a guide until he has more experience of his own.

Rule 1   If a game or experiment has a finite number of outcomes, and if one can envisage sequences of independent repetitions of the game or experiment, and if one is prepared to claim the sort of regularity for the proportions of occurrences of the outcomes exemplified above, then one can apply the theory of finite probability spaces to the game or experiment.

Rule 2 The elements of the sample space S of a finite probability space are to be identified with the outcomes of such a game or experiment which must be specified in such detail as to ensure that for each acceptable run or trial precisely one outcome occurs. Subsets of S correspond to events which are described by statements (such as, for example, the result of rolling a die is an even number) which limit the outcome without necessarily specifying it completely.

The function P of a finite probability space for application to a particular game may be chosen by means of proportions of occurrences of outcomes in a long sequence of runs of that game, and tested for validity by further sequences of runs of the game. Or the function may be chosen on the basis of previous experience and tested for validity by a sequence of runs of the game. For games which are played with symmetrical objects such as ordinary dice, in the absence of evidence to the contrary, one may assume that all the outcomes are equally likely and choose the 'equally-likely' model in which $\forall x, y \in S, P(\{x\}) = P(\{y\})$. Of course, if the die was a large one made of polystyrene with metal bolts for the pips, it would be highly unsymmetrical, and it would be unreasonable to choose the 'equally-likely' model for it. The third rule specifies terminology which will be used for the function P of a finite probability space which has been ascribed to a particular game. Note that there is a two way transfer of ideas between the game and the model. One starts with an assessment of whether one elementary event is more likely than another or whether they are equally likely, and one chooses probabilities for the elementary events. One then calculates the probabilities of compound events and uses the phrases more likely than and equally-likely for compound events as directed by the next rule.

Rule 3 For each event E, $P(E)$ is called the probability of E. If $P(E) > P(F)$, then E is said to be more likely than F. If $P(E) = P(F)$, then E and F are said to be equally likely.

The fourth rule concerns the application of finite probability spaces to two or more games played simultaneously. One can think of the game of casting a pair of dice as the combination of two games of casting a single die played simultaneously. With ordinary dice one expects that the way that one rolls will not affect the way that the other rolls. On the other hand if both the dice were made of strongly magnetized metal then when they were rolled together the interaction of the magnetic fields would affect the way in each of the dice rolled. We have an everyday sense of independence of simple games, and a scientist will have an everyday sense of independence of some of the experiments which he performs. In particular, he will have a sense of independence of repeated trials of an experiment, as is required in Rule 1 for the application of finite probability spaces to experiments. Let us consider now a composite game which consists of casting a die and tossing a coin independently of each other. If we take the usual outcomes for the die and for the coin, then the outcomes for the composite game can be described as

$$(1,h), \quad (2,h), \quad (3,h), \quad (4,h), \quad (5,h), \quad (6,h),$$
$$(1,t), \quad (2,t), \quad (3,t), \quad (4,t), \quad (5,t), \quad (6,t).$$

Again, if we suppose that the die and the coin are unbiased, then it is reasonable to suppose that in any sequence of runs of the game each of these outcomes will occur about one twelfth of the times. A reasonable model for this composite game is the equally-likely model. On the other hand, if we suspect that the die and the coin are biased, and have decided upon some probability function P_1 to represent the die and some probability function P_2 to represent the coin, then these can be used to obtain a probability function for the composite game. In any sequence of runs of the composite game the proportion of occurrences of the event '$(1, h)$ or $(1, t)$' will be approximately $P_1(\{1\})$, and the proportion of occurrences of the event $(1, h)$ will be approximately $P_1(\{1\}) \cdot P_2(\{h\})$, and similarly for the other elementary events. A reasonable probability function P for the composite game will be such that

$$P(\{(1,h)\}) = P_1(\{1\}) \cdot P_2(\{h\}),$$
$$P(\{(1,h),(3,t)\}) = P_1(\{1\}) \cdot P_2(\{h\}) + P_1(\{3\}) \cdot P_2(\{t\}), \quad \text{etc.}$$

That this procedure can be generalized to apply to the combination of any two games follows from the definition and proposition below.

Definition 5.2   *The cartesian product* of two finite probability spaces (S_1, P_1) and (S_2, P_2) is the ordered pair $(S_1 \times S_2, P)$ where the function

$$P:\mathscr{P}(S_1 \times S_2) \rightsquigarrow R$$

is defined by the equation

$$P(A) = \sum_{(x,\,y)\in A} P_1(\{x\}).P_2(\{y\}).$$

Note that for subsets of $S_1 \times S_2$ of the form $E \times F$, where $E \subseteq S_1$ and $F \subseteq S_2$,

$$P(E \times F) = P_1(E).P_2(F).$$

Proposition 5.1 The cartesian product of two finite probability spaces is a finite probability space.

Proof Suppose that (S_1, P_1) and (S_2, P_2) are finite probability spaces. We prove that $(S_1 \times S_2, P)$, as defined above, satisfies the five conditions of Definition 5.1.

(i) The number of elements in $S_1 \times S_2$ is the product of the number of elements in S_1 and the number of elements in S_2, and so is finite.

(ii) The domain of P is $\mathscr{P}(S_1 \times S_2)$. Moreover, since P_1 and P_2 are functions, if $P(A) = \lambda_1$ and $P(A) = \lambda_2$, then $\lambda_1 = \lambda_2$. Hence P is a function.

(iii) $P(S_1 \times S_2) = P_1(S_1).P_2(S_2) = 1$.

(iv) Since $\forall (x, y) \in S_1 \times S_2, P_1(\{x\}).P_2(\{y\}) \geqslant 0$,
   it follows that $\forall A \in \mathscr{P}(S_1 \times S_2), P(A) \geqslant 0$.

(v) If $A, B \subseteq S_1 \times S_2$ and $A \cap B = \varnothing$, then

$$P(A \cup B) = \sum_{(x,\,y)\in A \cup B} P_1(\{x\}).P_2(\{y\})$$

$$= \sum_{(x,\,y)\in A} P_1(\{x\}).P_2(\{y\}) + \sum_{(x,\,y)\in B} P_1(\{x\}).P_2(\{y\})$$

$$= P(A) + P(B).$$

In general, if one has two games to which have been ascribed finite probability spaces (S_1, P_1) and (S_2, P_2), then the outcomes of the composite game can be associated with the elements of the cartesian product space $S_1 \times S_2$. Moreover, if the games are thought to be independent in the everyday sense then it is reasonable and consistent to ascribe the cartesian product finite probability space $(S_1 \times S_2, P)$ to the composite game. The fourth rule for the use of

L

finite probability spaces as models for statistical situations specifies conditions for the use of the cartesian product.

Rule 4 If finite probability spaces have been ascribed to two games which are independent in the everyday sense, then the cartesian product of these two spaces is to be ascribed to the composite game which consists of the simultaneous performance of the two games.

The final rule concerns the role of certainty within a finite probability space model of a statistical experiment. For each probability space (S, P), the probability $P(S)$ of the certain event S is 1, and the probability $P(\emptyset)$ of the impossible event $\emptyset$ is 0. There may be other events which have the probability 1 or the probability 0. For example, one can envisage a finite probability space for the game of tossing a coin on to a table for which the elementary events are chosen to be symbols h, t, e, d which represent heads, tails, the coin balances on its edge on the table, and the coin drops off the table. Reasonable probabilities for these elementary events could be

$$P(\{h\}) = P(\{t\}) = \tfrac{2}{5}, \quad P(\{e\}) = 0, \quad P(\{d\}) = \tfrac{1}{5}.$$

Thus $P(\{t, h, d\}) = 1$. These express the belief that the coin is likely to drop off the table sometimes, but when it stays on the table it is likely to land on a side with heads and tails showing in roughly equal proportions. The probability of the theoretically possible event $\{e\}$ actually occurring is negligible. An event E such that $P(E) = 0$ will be said to be *effectively impossible* and an event F such that $P(E) = 1$ will be said to be *effectively certain*.

Rule 5 If an event is effectively certain, then its occurrence will be predicted with entire confidence.

As in Rule 3 there is here a two way transfer of ideas between the game and a probability model of the game. For simple games we start with assessments of the effectively certain events and build these ideas into the probability models of the games. Thus this rule is a reasonable expression of our ideas about elementary probability models of simple games. It becomes a powerful weapon when it is applied to effectively certain events in probability spaces such as those which arise through repeated use of Rule 4. It can be proved that for every probability space (S, P) and every element $x \in S$ it is effectively certain that in an infinite sequence of choices of elements from S the proportion of occurrences of x tends to $P(x)$ as a limit. Thus if you are prepared to ascribe a probability space (S, P) as a model of a game then you must predict with entire confidence that for every outcome x and every infinite sequence of runs of the game the proportion of occurrences of the outcome x tends to $P(x)$ as a limit. Since this is a theorem about infinite sequences of elements from S, it can only be proved in the setting of probability spaces with infinite sample spaces, which is beyond the scope of this book.

Exercises 5.2

1. Construct finite probability spaces to serve as models for the following situations.
   (a) A die is made of polystyrene and it has lead spheres embedded in it to serve as the spots. The game consists of casting the die once on a table.
   (b) The game consists of rolling a pair of dice on a table.
   (c) The game consists of opening the *Encyclopaedia Brittanica* at any page, choosing a word at random and recording the number of letters.
   (d) The game consists of choosing a football match played in the last ten years in the English Football League and recording the score.
   (e) The game consists of choosing a natural number less than 50 and recording the number of distinct divisors it possesses.
   (f) The game consists of choosing a person living in your town and recording whether or not he is taller than yourself.

2. Let (S, P) be a finite probability space.
   (a) Show that

   $$P(A^c) = 1 - P(A).$$

   (b) Show that

   $$P\left(\bigcup_{i=1}^{n} A_i\right) \leqslant \sum_{i=1}^{n} P(A_i).$$

   (c) Show that

   $$P\left(\bigcap_{i=1}^{n} A_i\right) \geqslant 1 - \sum_{i=1}^{n} P(A_i^c).$$

(d) Show that

$$P(S\backslash(A_1\cup A_2\cup A_3)) = 1 - P(A_1) - P(A_2) - P(A_3)$$
$$+ P(A_1\cap A_2) + P(A_2\cap A_3) + P(A_3\cap A_1)$$
$$- P(A_1\cap A_2\cap A_3).$$

Find a similar formula for

$$P\left(S\backslash\bigcup_{i=1}^{n} A_i\right).$$

5.3 CONDITIONAL PROBABILITIES

Suppose that a coin is to be tossed six times. What is the probability that if on the first five occasions it shows heads, then on the sixth occasion it shows heads? This is an example of a type of probability question which is met with in a wide range of applications of statistics. For example, in insurance one needs to know what is the probability that a man will reach the age of 65, given that he is 25 years old. What is the probability that if there are two faulty articles in a sample of 10 chosen from a batch of 100 then there are more than 15 faulty articles among the 100? These are all problems of determining the probability of one event F given that another event E has occurred.

In a sense all problems in probability are of this type. In allocating a finite probability model (S, P) to a game, we allocate a probability that the result of a run is the event F given that the result of the run is acceptable. One might reject as unacceptable, for example, those runs of a game for which a tossed penny is lost, and those runs of a chemical experiment for which the apparatus blows up. If one is only ever interested in those runs of a game in which the event E occurs, then runs in which E does not occur can be dismissed as unacceptable, and E can be taken as the sample space for a probability space model.

The more interesting problem is to relate a finite probability model (S, P) for a game to a finite probability model for statements of the kind 'the probability of F given E', where E and F are subsets of S. In this case every event $F \subseteq S$ gives rise to an event $E\cap F \subseteq E$, and a probability function $P_E: \mathscr{P}(E) \rightsquigarrow R$ can be defined in terms of the function $P: \mathscr{P}(S) \rightsquigarrow R$ in a way suggested by consideration of occurrences of events in a long sequence of runs of the game.

If in a sequence of runs of the game the number of occurrences of the elementary event $x \in S$ is denoted by $n(x)$, then the proportion of occurrences of a particular elementary event $x_0 \in E$ relative to the whole set S is $n(x_0)/\sum_{x \in S} n(x)$, while the proportion of occurrences of x_0 relative to the event E is $n(x_0)/\sum_{x \in E} n(x)$. Thus for the elementary events in E the proportions of occurrence relative to E are obtained from the proportions of occurrence relative to S by a linear scaling with ratio $\sum_{x \in S} n(x)/\sum_{x \in E} n(x)$. It is reasonable to model this in the probability theory by defining a function

$$P_E : \mathscr{P}(E) \rightsquigarrow R$$

which is obtained from the function $P:\mathscr{P}(S) \to R$ by a linear scaling which ensures that $P_E(E) = 1$.

The pair of equations

$$P_E(E \cap F) = k \, . \, P(E \cap F) \text{ and } P_E(E) = 1.$$

cannot be solved for k if $P(E) = 0$. If $P(E) \neq 0$, then $k = 1/P(E)$, and $P_E(E \cap F) = P(E \cap F)/P(E)$. The condition $P(E) \neq 0$ is not restrictive in practice since events E which are effectively impossible do not give rise to problems of the type discussed above.

In statistics one traditionally uses the notation

$$P(F \,|\, E) = P_E(E \cap F) = P(E \cap F)/P(E).$$

The event E is called the *conditioning event* and $P(F \,|\, E)$ is read 'the *conditional probability* of F given E'.

The conditional probability of E given F and the conditional probability of F given E are closely related. We have

$$P(E \,|\, F) = \frac{P(F \cap E)}{P(F)} = \frac{P(F \,|\, E) \, . \, P(E)}{P(F)}.$$

This is one form of a statement which is known in statistics as Bayes' Theorem. A more general form of the theorem arises when one relates a single event F to a family of events $E_1, E_2, \ldots, E_n$

which partition S. The events E_i, $i = 1, 2, \ldots, n$, are then pairwise disjoint subsets of S whose union is S, so that $P(E_1) + P(E_2) + \ldots + P(E_n) = 1$.

Proposition 5.2 *Bayes' Theorem*   If (S, P) is a finite probability space and $E_1, \ldots, E_n$ is a partition of S, and F is a subset of S such that
$$P(E_1) \neq 0, \ldots, P(E_n) \neq 0, P(F) \neq 0,$$
then

$$P(E_i \,|\, F) = \frac{P(F \,|\, E_i) \cdot P(E_i)}{\sum\limits_{j=1}^{n} P(F \,|\, E_j) \cdot P(E_j)}$$

for $i = 1, 2, \ldots, n$.

Proof   By definition,

$$\sum_{j=1}^{n} P(F \,|\, E_j) \cdot P(E_j) = \sum_{j=1}^{n} P(E_j \cap F).$$

Now $\bigcap\limits_{j=1}^{n} (E_j \cap F) = \varnothing$, and by exercise 6 of §3.8 $\bigcup\limits_{j=1}^{n} (E_j \cap F) = F$.

Hence by the additivity of the function P, $\sum\limits_{j=1}^{n} P(E_j \cap F) = P(F)$.

Thus $\dfrac{P(F \,|\, E_i) \cdot P(E_i)}{\sum\limits_{j=1}^{n} P(F \,|\, E_j) \cdot P(E_j)} = \dfrac{P(E_i \cap F)}{P(F)} = P(E_i \,|\, F)$.

In the discussion of the use of probability spaces as models of statistical situations in the last section we considered games which were thought to be independent in the everyday sense. We also have an everyday sense of independence of two events, which is that knowledge of the occurrence of one of the events gives no information about the occurrence of the other event. Corresponding to this everyday notion about a statistical situation there is a defined concept concerning probability models which we arrive at by considering the formal statements $P(E \,|\, F) = P(E)$ and $P(F \,|\, E) = P(F)$. These conditions hold if and only if $P(E) \neq 0$, $P(F) \neq 0$ and $P(E \cap F) = P(E) \cdot P(F)$.

We drop the restrictions that E and F must not be effectively impossible and define two events E and F to be *independent in* (S, P) if $P(E \cap F) = P(E) \cdot P(F)$. The relationship between independence of actual events in the everyday sense, and formal independence of conceptual events within a probability space model is delicate, and which one of them is dominant at any time depends on the particular statistical problem under consideration. We note however that the definition of independence of events in a probability space is consistent with the Rule 4 which specifies the use of the cartesian product probability space for a composite game consisting of the simultaneous performance of two games which are thought to be independent in the everyday sense. Suppose that (S_1, P_1) and (S_2, P_2) are the probability spaces ascribed to the two separate games. Then the product probability space $(S_1 \times S_2, P)$ must be ascribed to the composite game. Two events $E_1 \subseteq S_1$ and $E_2 \subseteq S_2$ give rise to events $E_1 \times S_2 \subseteq S_1 \times S_2$ and $S_1 \times E_2 \subseteq S_1 \times S_2$. If we know that the event $E_1 \times S_2$ occurs in the composite game then we know that the outcome of the first game is restricted to belong to E_1 but we know of no restriction on the outcome of the second game, so we know nothing about the occurrence of the event $S_1 \times E_2$. The events $E_1 \times S_2$ and $S_1 \times E_2$ are independent in the everyday sense. That they are also independent in $(S_1 \times S_2, P)$ in the sense defined above follows from the calculation

$$P(E_1 \times S_2 \cap S_1 \times E_2) = P(E_1 \times E_2) = P_1(E_1) \cdot P_2(E_2).$$

Bayes' theorem can be applied to statistical problems in a wide variety of fields. The method of application can be illustrated by the following problem concerning three urns containing coloured balls. Urn U_1 contains 1 red, 2 white and 2 blue balls; urn U_2 contains 3 red, 2 white and 1 blue balls; urn U_3 contains 1 red, 1 white and 2 blue balls. A ball is chosen at random from one of the urns. What is the probability that the ball came from U_1, given that the ball is red?

There is no absolute answer to this question. It can only be answered in relation to a finite probability model (S, P) for the situation. A suitable sample space S could be the cartesian product $\{U_1, U_2, U_3\} \times \{r, w, b\}$, in which a typical elementary event (U_1, r) would represent the choice of a red ball from the first urn. The choice of any ball from the first urn would then be represented by the event $\{U_1\} \times \{r, w, b\} \subseteq S$. Where no confusion will result, this long notation will be replaced by U_1. Thus $P(U_1)$ will be used instead of $P(\{U_1\} \times \{r, w, b\})$ to denote the probability that a ball has been chosen from the first urn. Similarly, $P(r)$ will be used instead of $P(\{U_1, U_2, U_3\} \times \{r\})$ to denote the probability that a red ball has been chosen from some urn. Corresponding abbreviations will be used for conditional probabilities. For example, the required probability that the ball came from U_1, given that the ball is red will be denoted by $P(U_1 \mid r)$.

It is reasonable to suppose that the three urns are equally likely to be chosen, and that within each urn the balls are equally likely to be chosen. Thus a reasonable probability function for the problem has $P(U_1) = P(U_2) = P(U_3) = \frac{1}{3}$, $P(r \mid U_1) = \frac{1}{5}$, $P(r \mid U_2) = \frac{1}{2}$, $P(r \mid U_4) = \frac{1}{4}$. The conditional probabilities of white and blue balls, given the urns from which the balls came, can also be determined by the criterion of equal likelihood. However, they are not needed to determine the required conditional probability $P(U_1 \mid r)$. According to Bayes' theorem,

$$P(U_1 \mid r) = \frac{P(r \mid U_1) . P(U_1)}{\sum\limits_{j-1}^{3} P(r \mid U_j) . P(U_j)}$$

$$= \frac{\frac{1}{5} \cdot \frac{1}{3}}{\frac{1}{5} \cdot \frac{1}{3} + \frac{1}{2} \cdot \frac{1}{3} + \frac{1}{4} \cdot \frac{1}{3}} = \frac{4}{19}.$$

This then is the answer to the question on the basis of the chosen model.

Exercises 5.3

1. Show that $P(E \mid F) = P(E)$ and $P(F \mid E) = P(F)$ if and only if $P(E) \neq 0$, $P(F) \neq 0$ and $P(E \cap F) = P(E) . P(F)$.

2. Consider the following questions in relation to the models you have constructed for the situations described in exercise 1 of §5.2.

 (a) What is the probability that the weighted die shows 2, given that it shows an even number?
 (b) What is the probability that the two dice show the same number, given that the total shown is 8?

(c) What is the probability that the chosen word has ten letters given that it has six or more letters?

(d) Given that a team scores 4 goals, what is the probability of its winning the match?

(e) Given that a natural number less than 50 has an odd number of divisors, what is the probability that it has exactly 5 or exactly 7 divisors?

(f) Given that a person is taller than yourself, what is the probability that he is over 6 ft. tall?

3. Of two similar coins one is unbiased and the other is biased with $p(h) = \frac{3}{4}$. A coin is selected at random and tossed. Assuming that the two coins are equally likely to be chosen determine the following probabilities.

(a) The probability that a head occurs when the coin is tossed once.

(b) The conditional probability that the selected coin is biased, given that a head occurs when the coin is tossed once.

(c) The conditional probability that the selected coin is biased, if two heads occur when the coin is tossed twice.

4. It is proposed that a swimming pool should be built in a town in which 30% of the population are under the age of 18, 50% are between 18 and 65 while the rest are over 65. Of those under 18, 85% are in favour of the pool; of those between 18 and 65, 60% are in favour; while of those over 65 only 40% are in favour. What is the conditional probability that a person is under 18, given that he is in favour of the pool?

5. Of four machines which produce the same item, three of them have the same capacity and the fourth has twice that capacity. The probabilities that an item from the first three machines is defective are 0·04, 0·1 and 0·05 respectively. The probability that an item from the fourth machine is defective is 0·2. What is the conditional probability that a defective item was produced by the fourth machine.

6. Assuming that cards are dealt at random, what is the probability that a hand of cards contains no trumps? If a whist player observes that he has six trump cards in his own hand, what in his view is the conditional probability that his partner has no trumps?

5.4 TOSSING COINS

We now return to the question: what is the probability that if a coin is tossed five times and shows heads each time, then it will show a head the next time it is tossed? The question can only be answered in relation to a probability space model for the coin.

There are several ways to view the problem. You may decide to treat the coin as an 'ordinary' one, and to ascribe to it the equally-likely probability model on the basis of past experience of similar coins and of arguments about the symmetry of the coin. On the other hand you may decide to keep an open mind as to whether or not the coin has been tampered with to make it biased, and try to determine a reasonable probability function to ascribe to it. In this case you may decide to toss the coin many times and use the frequencies of occurrences of heads and tails to ascribe a probability model to the coin. Alternatively you may decide to use the limited information available from tossing the coin five times to make some limited statement about the possible extent of the bias.

Suppose that you have ascribed some probability model to the game which consists of 'toss a coin and record h or t', and that you wish to use this to determine the probability that a sixth head occurs after a run of five heads. Then you will have decided on two numbers $p(h)$ and $p(t)$ whose sum is 1. According to Rule 4, for the game which consists of 'toss the coin six times and record the results in order' the cartesian product probability space must be used. For this the sample space is

$$S = \{(x_1, \ldots, x_6) \,|\, \forall i, x_i = h \text{ or } x_i = t\}$$

which has 64 elements, and the probability function is defined by

$$P(\{(x_1, \ldots, x_6)\}) = p(x_1) \cdot p(x_2) \ldots p(x_6).$$

The conditioning event is

$$E = \{(h, h, h, h, h, h)\} \cup \{(h, h, h, h, h, t)\}$$

and

$$P(E) = p(h)^6 + p(h)^5 \cdot p(t) = p(h)^5.$$

The event in which we are interested is

$$F = \{(x_1, \ldots, x_6) \,|\, x_6 = h\}.$$

Thus $E \cap F$ has just the one element (h, h, h, h, h, h) and $P(E \cap F) = p(h)^6$. It follows that $P(E|F) = p(h)$. In particular, if the equally-likely model has been ascribed, then the probability that a sixth head occurs, given that there has been a run of five heads, is $\frac{1}{2}$.

In the above approach to the problem we declare ourselves effectively certain about a probability model for a single toss of the coin and on the basis of that model make a firm statement about the conditional probability. There is an implication that the model has been chosen and must remain fixed whatever the evidence of a sequence of tosses of the coin. However, a more flexible approach to the problem is to choose an initial model, and to use it just so long as the evidence of a sequence of tosses is reasonably acceptable within that model. On the basis of the equally-likely model, the probability of a run of four heads is $(\frac{1}{2})^4 \simeq 0.06$, which is reasonably acceptable. That of a run of five heads is $(\frac{1}{2})^5 \simeq 0.03$, which is getting small. The probability of a run of six heads is 0.015 which may make one begin to doubt that the coin is unbiased. The probability of a run of seven heads is 0.008. This is so small that a run of seven heads may easily make you decide to try another model which reflects the apparent bias towards heads, say $p(h) = \frac{3}{4}$, $p(t) = \frac{1}{4}$. On the basis of this second model, the probability of a run of seven heads is $(\frac{3}{4})^7 \simeq 0.13$, which is acceptable, while that of a run of ten heads is 0.056 which is getting small. The probability of a run of 16 heads is 0.01, and so a run of 16 heads may make you decide to reject the second model in favour of a third model which expresses an even greater bias towards heads.

An alternative approach to the problem is one in which we admit that we are uncertain as to what probability function to use for a single toss of the coin and hope to get some guidance from a few tosses of the coin together with a more complicated probability model for the situation. In order to set up this model, one must make some subjective judgements about the situation in which coin tossing is taking place. One may judge that the possibility of bias has to be taken into account because there is evidence that the coin has been tampered with, and one may judge from the appearance of the coin that it is probably biased towards heads. One may judge that there is a possibility of bias because the coin has been produced by a gambler who is known to use loaded dice and coins. A method of analysing the situation using Bayes' Theorem can be illustrated by an imaginary situation in which a gambler has a set of coins of which he knows the bias, from which he chooses one for his victim to toss. While the gambler knows the degrees of bias of the coins he has available, his victim can only assess what is likely. Having handled the coin and assessed the situation, the victim must decide on some more details. We shall suppose that he has decided that the gambler is likely to have four coins for which the probabilities are given by

$$p_4(h) = 0·4, \quad p_5(h) = 0·5, \quad p_6(h) = 0·6, \quad p_7(h) = 0·7,$$

respectively, and that the gambler chooses between them with equal probability. Then the victim has decided that the total result of the game 'take a coin from the gambler, toss it and record heads or tails' is to be one of the letters $\{h, t\}$, together with one of the parameters $\{4, 5, 6, 7\}$. While the gambler knows both the letter and the parameter, the victim knows only the letter, from which he hopes to gain some information about the parameter.

The rather complicated probability model for the game chosen by the victim can be presented formally as follows. The sample space is $S \times A$, where $S = \{h, t\}$ and $A = \{4, 5, 6, 7\}$. The probability function $P : \mathscr{P}(S \times A) \rightsquigarrow R$ is determined by the conditional probabilities of heads given the parameter $a \in A$ appropriate for the coin, and by the a priori assessment that the four parameters are equally likely to occur. Thus

$$P(\{h\} \times A \,|\, S \times \{4\}) = p_4(h) = 0·4,$$

$$P(\{t\} \times A \,|\, S \times \{4\}) = 1 - p_4(h) = 0·6,$$

and similarly for the other three parameters. Also

$$P(S \times \{4\}) = P(S \times \{5\}) = P(S \times \{6\}) = P(S \times \{7\}) = 0·25.$$

It follows that, for $a \in A$,

$$P(h, a) = 0·25 . p_a(h) \text{ and } P(t, a) = 0·25(1 - p_a(h)).$$

According to this a priori probability model, the probability that a head will show is

$$P(\{h\} \times A) = P(h, 4) + P(h, 5) + P(h, 6) + P(h, 7)$$

$$= \sum_{a=4}^{7} p_a(h) . 0·25 = 0·55.$$

This reflects the victim's subjective assessment before tossing the coin that it is likely to be slightly biased towards heads.

If on tossing the coin a head shows, this will be confirming evidence of bias towards heads. A numerical value can be placed on the evidence if one considers the conditional probabilities of the parameters given that a head has turned up, namely $P(S \times \{a\} \mid \{h\} \times A)$ for $a \in A$. By Bayes' theorem we have

$$P(S \times \{4\} \mid \{h\} \times A) = \frac{P(\{h\} \times A \mid S + \{4\}) \cdot P(S \times \{4\})}{\sum\limits_{a=4}^{7} P(\{h\} \times A \mid S \times \{a\}) \cdot P(S \times \{a\})}$$

$$= \frac{0 \cdot 4 \times 0 \cdot 25}{2 \cdot 2 \times 0 \cdot 25} \simeq 0 \cdot 182.$$

Similarly

$$P(S \times \{5\} \mid \{h\} \times A) \simeq 0 \cdot 227,$$

$$P(S \times \{6\} \mid \{h\} \times A) \simeq 0 \cdot 273,$$

$$P(S \times \{7\} \mid \{h\} \times A) \simeq 0 \cdot 318.$$

These conditional probabilities can be used to ascribe a new probability function P^* to $S \times A$ called the *a posteriori* probability function, in which

$$P^*(S \times \{a\}) = P(S \times \{a\} \mid \{h\} \times A), \quad a \in A.$$

In this *a posteriori* model, the probability that a head will show on tossing a coin is

$$P^*(\{h\} \times A) = P^*(h, 4) + P^*(h, 5) + P^*(h, 6) + P^*(h, 7)$$

$$= 0 \cdot 4 P^*(S \times \{4\}) + 0 \cdot 5 P^*(S \times \{5\})$$
$$+ 0 \cdot 6 P^*(S \times \{6\}) + 0 \cdot 7 P^*(S \times \{7\})$$

$$\simeq 0 \cdot 4 \times 0 \cdot 182 + 0 \cdot 5 \times 0 \cdot 227 + 0 \cdot 6 \times 0 \cdot 273 + 0 \cdot 7 \times 0 \cdot 318$$

$$\simeq 0 \cdot 573$$

This is only slightly different from the probability ascribed by the *a priori* model. However, if we now perform the corresponding calculation for the case in which the coin has been tossed five times and five heads have occurred, we shall see that the evidence for bias is then much stronger. In this case one must analyse *a priori* and *a posteriori* probabilities on the space $S \times S \times S \times S \times S \times A$. The full notations in this case are long and so will be abbreviated: $S \times S \times S \times S \times S \times \{a\}$ will be reduced to a and $(h, h, h, h, h) \times A$ will be reduced to h^5. The *a priori* probability function will be such that $P(a) = 0.25$, $a \in A$. After five heads have occurred, the *a posteriori* probability function will be determined by

$$P^*(a) = P(a \,|\, h^5) = \frac{P(h^5 \,|\, a) . P(a)}{\sum\limits_{a=4}^{7} P(h^5 \,|\, a) . P(a)}.$$

Now $P(h^5 \,|\, 4)$ is obtained by fixing the parameter value at 4 so that the probability of a head is $p_4(h) = 0.4$, and therefore the probability of five heads is $(0.4)^5$. By similar calculations, it follows that

$$P^*(4) = \frac{(0.4)^5}{(0.4)^5 + (0.5)^5 + (0.6)^5 + (0.7)^5} \simeq 0.0356,$$

and also that $P^*(5) \simeq 0.1088$, $P^*(6) \simeq 0.2706$, $P^*(7) \simeq 0.5850$. In this probability model, the probability that on the next throw another head will turn up is

$$\sum_{a=4}^{7} p_a(h) . P^*(a) \simeq 0.64,$$

which is considerably larger than the *a priori* probability, as is to be expected.

You will see that this method of using Bayes' theorem to wrest conclusions from short runs of statistical information depends upon many subjective judgements in the process of building the *a priori* model. In much of statistics, an important part of the work of the statistician is to help to decide which type of model can be used to analyse the situation in hand. In this he must work closely with the person who presents the problem and who will use the results of the statistical analysis. For like all conceptual models in mathematics, a statistical model must be simple enough for it to be possible to calculate within the model, and also fit the situation sufficiently well so that the results of the calculations can be applied with confidence.

SOLUTIONS AND ANSWERS
TO SELECTED EXERCISES

Exercises 1.2

2. The following are true: (a), (d), (e), (h), (j), (l).

3. (a) $\{1,2,3,4,5,6,7,8,9\}$ or $\{1,2,3,\ldots,9\}$.
   (b) $\{2^n \mid n \in N \text{ and } 1 \leqslant n \leqslant 5\}$.
   (c) $\{x \mid x \in R \text{ and } x^2 - 3x + 2 = 0\} = \{1,2\}$.
   (d) $\{a_i \mid i \in N \text{ and } i \leqslant 20\} = \{a_1, a_2, a_3, \ldots, a_{20}\}$.

4. (a) $\{3\}$. (b) $\{4,5\}$. (c) $\{1, 2, 3, 4, \ldots\}$.

5. There are no natural numbers x for which $x^2 - x + 6 = 0$. One could decide to denote this by a pair of braces with nothing between, $\{\ \}$. However, the special symbol $\varnothing$ will be introduced in §1.3 to deal with situations such as this.

6. The first four diagrams yield $f(1) = 1$, $f(2) = 2$, $f(3) = 4$, $f(4) = 8$. One might reasonably predict from these that $f(5) = 16$, $f(6) = 32$ and in general $f(n) = 2^{n-1}$. However, this prediction is incorrect. In fact $f(5) = 16$, $f(6) = 31$ and in general

$$f(n) = 1 + \binom{n}{2} + \binom{n}{4}.$$

Exercises 1.3

1. $A \cup B = \{1,2,3,4,5,6,8\}$, $A \cap B = \{2,4\}$, $A \backslash B = \{6,8\}$,
   $A \triangle B = \{1,3,5,6,8\}$, $A^c = \{1,3,5,7\}$, $(B \backslash A)^c = \{2,4,6,7,8\}$.

2. $A \cup B = \{x \mid x \in R \text{ and } 0 < x \leqslant 3\}$, $A \cap B = \varnothing$, $A \backslash B = A$,
   $B \backslash A = B$, $A \triangle B = A \cup B$.

3. $A \cup B = A$, A^c is the set of odd integers, B^c is the set of integers which are not powers of two.

4. $Z \cap N = N$, $Q \cup N = Q$, $R \backslash Q$ is the set of irrational numbers, $R \triangle N = R \backslash N$, and is the set of those real numbers which are not positive integers.

5. The exercise anticipates the work of the next two sections. The set $X = A \cup B$ satisfies the required equations as does every larger set. In particular, if A and B are chosen from a universal set U, then $X = U$ satisfies the equations. The set $Y = A \cap B$

satisfies the required equations, as does every smaller set. In particular, $Y = \varnothing$ satisfies the equations.

Exercises 1.4

1. The following are true: $C \subseteq B, C \subseteq A, C \subset A, C \subset B, \varnothing \subset B$.
2. The following are true: $Q \subseteq R, Q \subset R, N \subseteq Z, Z \subseteq R$.
3. Statement (a) is true. Statement (b) is false.
4. $\{\varnothing, \{a\}, \{b\}, \{c\}, \{d\}, \{a,b\}, \{a,c\}, \{a,d\}, \{b,c\}, \{b,d\}, \{c,d\}, \{a,b,c\}, \{a,b,d\}, \{a,c,d\}, \{b,c,d\}, A\}$.
5. $\{\varnothing, \{a\}, \{b\}, \{\{a\}\}, \{a,b\}, \{a,\{a\}\}, \{b,\{a\}\}, A\}$.
6. The following are true: $C \in A, \; C \subset B, \; C \in \mathscr{P}(B), \; \varnothing \in \mathscr{P}(A), \; \varnothing \subseteq \mathscr{P}(A), \{\varnothing\} \subseteq \mathscr{P}(A)$.
7. 2^n.

Exercises 1.5

2. (a) $\cap$.   (b) $\backslash$.   (c) $\cup$.   (d) $\cap$.

Exercises 2.3

1. (a) $[$(attach rope) $\wedge$ (release brake) $\wedge$ (pull hard)$]$
        $\Rightarrow [$(lorry moves) $\vee$ (rope breaks)$]$.
   (b) $[$(feel drowsy) $\wedge$ (driving)$]$
        $\Rightarrow [$(stop and rest) $\vee$ (open window and admit fresh air)$]$.
   (c) $(ax^2 + bx + c = 0$ has real roots$) \Rightarrow (b^2 - 4ac \geqslant 0)$.
   (d) (travel worthwhile)
        $\Rightarrow [$(try local food) $\wedge$ (investigate local customs)$]$.
2. (a) I work hard and I am happy.
   (b) If I work hard then I enjoy mathematics or I am happy.
   (c) I am unhappy if and only if I work hard and don't enjoy mathematics.
   (d) I enjoy mathematics and I work hard or I am happy.
   (e) If I enjoy mathematics then I work hard or I am unhappy.

Exercises 2.4

1.

p	q	(a)	(b)	(c)	(d)
T	T	T	T	T	T
T	F	T	T	F	T
F	T	T	F	F	T
F	F	T	F	F	T

(a), (d), (f) and (h) are tautologies.

4.

p	q	either p or q
T	T	F
T	F	T
F	T	T
F	F	F

The statement $(p \vee q) \wedge [\sim(p \wedge q)]$ has the same truth table. So has the statement $\sim p \Leftrightarrow q$.

5. The statement $(\sim p \vee q) \wedge (\sim q \vee p)$ has the same truth table as $p \Leftrightarrow q$.

6. $\sim q$ is equivalent to $q|q$, and $p \wedge q$ is equivalent to $(p|q)|(p|q)$. The connectives $\vee$, $\Leftrightarrow$ and $\Rightarrow$ can be expressed in terms of $\sim$ and $\wedge$ and so in terms of $|$.

Exercises 2.5

1. (a) $\forall x, \quad x \in Z \Rightarrow x \in R$.
   (c) $\exists x \in Q, \quad x < 0$.
   (e) $\forall x, \quad x \in Z \Rightarrow \cos \pi x = 1$.
   Statements (a), (c), (d) are true.

2. (a) Every integer is greater than 5 and less than 10.
   (c) The numbers 1, 2, 3 are all roots of the equation
   $x^3 - 6x^2 + 11x - 6 = 0$.
   (e) The square of every real root of the equation $x^2 + 2x + 1 = 0$
   is equal to 1.
   Statements (b), (c), (e) and (f) are true.

3. (b) N.   (c) Q, R.

4. All except (a) and the reverse of (d) are normally taken to be tautologies.

M

5. Write $N_n = \{i \mid i \in N \wedge i \leqslant n\}$.

Then $\bigcup_{i=1}^{n} A_i = \{x \mid \exists i \in N_n, x \in A_i\}$,

and $\bigcap_{i=1}^{n} A_i = \{x \mid \forall i \in N_n, x \in A_i\}$.

Similarly $\bigcap_{i=1}^{\infty} A_i = \{x \mid \forall i \in N, x \in A_i\}$.

6. (b) $\forall x \in N, x^2 + x - 1 \neq 0$.   (d) $\exists x \in R, (x^2 > 4) \wedge (x \leqslant 2)$.

Exercises 2.6

1. (a) $\exists n \in Z, \exists m \in Z, 2n+1 = 2^m$.   (c) $\forall x \in R \backslash Q, \exists y \in Z, x < y < x+1$.
   Statements (a), (b), (c) are true.

2. (a) For every real number x the equation $x^2 - 5xy + 6y^2 = 0$ in
   y has a real root.
   (c) If y is a non-zero real number and x is a real number less than
   y then x/y is less than 1.
   Statements (a) and (d) are true.

4. Yes.

5. (a) $\forall x, \exists y, x \geqslant y$.
   (c) $\exists x, \exists y, \forall z, (xy < z^2) \vee (xy > z^2)$.

6. (a) True with $X \subseteq A$, so the solution is not unique unless $A = \varnothing$.
   (b) True with $X = \varnothing$, and the solution is unique.
   (c) True with $X \supseteq A$, so the solution is not unique unless $A = U$.
   (d) True with $X = U$, and the solution is unique.
   (e), (f) True with unique solution $x = 0$.
   (g), (h) True with unique solution $x = 1$.

7. As is exercise 6, for each equation there are two possible statements.
   (a) $\forall A \subseteq U, \exists X \subseteq U, A \cup X = U$ is true: take X such that
   $X \supseteq A^c$.
   (b) $\exists X \subseteq U, \forall A \subseteq U, A \cup X = U$ is true: $X = U$ is unique
   solution.
   (c) $\forall A \subseteq U, \exists X \subseteq U, A \cap X = \varnothing$ is true: take X such that
   $X \subseteq A^c$.
   (d) $\exists X \subseteq U, \forall A \subseteq U, A \cap X = \varnothing$ is true: $X = \varnothing$ is unique
   solution.

Exercises 2.7

1.

(a) 1. $A \subseteq A \cup B$ for $\forall x, x \in A \Rightarrow x \in A \cup B$ (I)

    2. $A \subseteq A \cup B$ for $\forall x, x \in A \Rightarrow x \in A \vee x \in B$ (III)

    3. $\forall x, x \in A \Rightarrow (x \in A \vee x \in B)$ (tautology vi)

    4. $A \subseteq A \cup B$.   (2 and 3)

(h) 1. $A \cap (B \backslash A) = \varnothing$ for $\forall x, \sim(x \in A \cap (B \backslash A))$ (V)

    2. $A \cap (B \backslash A) = \varnothing$ for $\forall x, \sim(x \in A \wedge x \in B \backslash A)$ (IV)

    3. $A \cap (B \backslash A) = \varnothing$ for $\forall x, \sim(x \in A \wedge x \in B \wedge \sim x \in A)$ (Def. of \\)

    4. $A \cap (B \backslash A) = \varnothing$ for $\forall x, \sim x \in A \vee \sim x \in B \vee x \in A$ (tautology xv)

    5. $\forall x, x \in A \vee \sim x \in A$ (tautology xii)

    6. $\forall x, \sim x \in A \vee \sim x \in B \vee x \in A$ (tautologies i and vi)

    7. $A \cap (B \backslash A) = \varnothing$ (4 and 6)

2.

(a) 1. $A \cap B = \varnothing$ for $\forall x, \sim(x \in A \wedge x \in B)$ (V)

    2. $(\forall x, \sim(x \in A \wedge x \in B)) \Rightarrow (\forall x, \sim x \in A \vee \sim x \in B)$
                         (de Morgan's law, tautology xv)

    3. $(\forall x, \sim x \in A \vee \sim x \in B) \Rightarrow (\forall x, x \in A \Rightarrow \sim x \in B)$
                              (tautology xiii)

    4. $(\forall x, x \in A \Rightarrow \sim x \in B) \Rightarrow (\forall x, (x \in A \wedge x \in A) \Rightarrow (x \in A \wedge \sim x \in B))$
                              (tautology xi)

    5. $(\forall x, (x \in A \wedge x \in A) \Rightarrow (x \in A \wedge \sim x \in B))$
        $\Rightarrow (\forall x, x \in A \Rightarrow (x \in A \wedge \sim x \in B))$ (tautology iv)

    6. $A \cap B = \varnothing \Rightarrow (\forall x, x \in A \Rightarrow (x \in A \wedge \sim x \in B))$
                      (from 1 to 5, using transitivity)

    7. $A \cap B = \varnothing \Rightarrow A \subseteq A \backslash B$ (from 6, using I and definition of \\)

    8. $\forall x, (x \in A \wedge \sim x \in B) \Rightarrow x \in A$ (tautology vii)

    9. $A \backslash B \subseteq A$ (from 8, using I and definition of \\)

    10. $A \cap B = \varnothing \Rightarrow A \backslash B \subseteq A$ (from 9, using tautology vi)

    11. $A \cap B = \varnothing \Rightarrow (A \subseteq A \backslash B \wedge A \backslash B \subseteq A)$
                      (from 7 and 10, using tautology x)

    12. $A \cap B = \varnothing \Rightarrow A = A \backslash B$ (from 11, using II)

(e) 1. $A \subseteq B \wedge B \subseteq C$ for $\forall x, x \in A \Rightarrow x \in B \wedge \forall x, x \in B \Rightarrow x \in C$
                                    (I)

    2. $(\forall x, x \in A \Rightarrow x \in B \wedge \forall x, x \in B \Rightarrow x \in C)$
       $\Rightarrow (\forall x, (x \in A \Rightarrow x \in B) \wedge (x \in B \Rightarrow x \in C))$
                          (see exercise 4 of §2.5)

3. $(\forall x, (x \in A \Rightarrow x \in B) \wedge (x \in B \Rightarrow x \in C)) \Rightarrow$
   $(\forall x, x \in A \Rightarrow x \in C)$ (tautology ix)
4. $(A \subseteq B \wedge B \subseteq C) \Rightarrow (\forall x, x \in A \Rightarrow x \in C)$
                  (from 1, 2 and 3, using transitivity)
5. $(A \subseteq B \wedge B \subseteq C) \Rightarrow A \subseteq C$ (from 4, using I)

3. The following converse statements are true.
   (a) If $A \backslash B = A$ then $A \cap B = \varnothing$.
   (b) If $A \backslash B = \varnothing$ then $A \subseteq B$.
   (d) If $B \subseteq A$ the $A^c \subseteq B^c$.

4. The first statement is proved directly. The second statement is proved by *reductio ad absurdum*—if x is rational and $x+y$ is rational then y is rational.

5. p for $x^2 = 1/\sqrt{2}$; q for 'x is irrational'. The proof proceeds as that of Proposition 2.7; or the result may be deduced from that proposition. The converse statement is false.

7. $\forall x, (\forall \varepsilon > 0, x < \varepsilon) \Rightarrow x \leqslant 0$ is equivalent to $\forall x, x > 0 \Rightarrow$ $\sim (\forall \varepsilon > 0, x < \varepsilon)$ by the law of contrapositive and is in turn equivalent to $\forall x, x > 0 \Rightarrow \exists \varepsilon > 0, x \geqslant \varepsilon$. This is a true statement, for if x_0 is arbitrary, then if $x_0 > 0$, $x_0 \geqslant x_0$, and so $\varepsilon = x_0$ satisfies $x_0 \geqslant \varepsilon$. The properties of real numbers used are those relating to inequality.

8. The tautology $(p \Rightarrow (q \vee r)) \Rightarrow ((p \wedge \sim q) \Rightarrow r)$ can be obtained from a truth table or from tautologies (xiii), (i), (v), (xv) and (ix). A statement of the form $(p \wedge \sim q) \Rightarrow r$ can be proved by the methods illustrated.

Exercises 3.2

1.
(e)

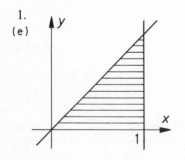

(h)

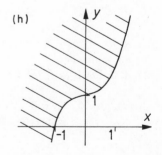

2. $A \times A = \{(a,a),(a,b),(b,a),(b,b)\}$.
   $A \times B = \{(a,c),(b,c),(a,d),(b,d),(a,e),(b,e)\}$.

3.

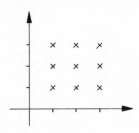

4. $m \times n$.

7. The elements of $(A \times B) \times C$ are of the form $((a_i, b_j), c_k)$ while the elements of $A \times (B \times C)$ are of the form $(a_i, (b_j, c_k))$. The two sets are not equal, but they have the same number of elements and there is a natural relationship between them which makes $((a_i, b_j), c_k)$ correspond to $(a_i, (b_j, c_k))$.

Exercises 3.3

1. $T_1^{-1} = \{(x,y) \mid x = 2, y = 3\}$. $T_3^{-1} = \{(x,y) \mid y \leqslant x\}$.
   $T_5^{-1} = \{(x,y) \mid y \leqslant x \text{ and } 0 \leqslant y \leqslant 1\}$.
   $T_7^{-1} = \{(x,y) \mid y^2 + x^2 \geqslant 3\}$.

2.

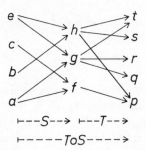

$T \circ S = \{(a,p), (a,q), (a,r), (a,t), (b,p), (b,s),$
$\qquad (b,t), (c,p), (e,p), (e,q), (e,r), (e,s), (e,t)\}$

3. $T \circ S = \{(x,y) \mid y = 2(x+1)^3 - 2\}$, $S \circ T = \{(x,y) \mid y = (2x-1)^3\}$.

4. $T_1 \circ T_3 = \{(x,y) \mid x^4 = 3, y = 2\}$, $T_2 \circ T_3 = \{(x,y) \mid x^4 = 2\}$.
   $T_2 \circ T_4 = \{(x,y) \mid x \leqslant 2\}$, $T_4 \circ T_8 = \{(x,y) \mid y > x^3 + 1\}$.

5. $T \circ S = \{(x, y) \mid y = ACx + BC + D\}.$
   $T \circ S = S \circ T$ if and only if $BC + D = AD + B.$

Exercises 3.4

1. (a) $A/S = \{A_0, A_1, \ldots, A_{m-1}\}$ where $\forall i,\ A_i = \{i + km \mid k \in Z\}.$
   (b) The equivalence relation S gives rise to a partition of A of which a typical element is

   $$\left\{ \frac{m}{2m} \,\middle|\, m \in Z \backslash \{0\} \right\} \subseteq A.$$

   This subset can be identified with the rational number 'one half'. Thus the quotient set A/S can be identified with the set Q of rational numbers. In books which treat the development of real numbers from an algebraic standpoint it is shown that the operations of addition and multiplication on Z can be used to define addition and multiplication on A/S.
   (c) The equivalence relation S gives rise to a partition of A of which typical subsets are $\{(n, n+2) \mid n \in N\}$ and $\{(n+2, n) \mid n \in N\}$. These subsets can be identified with the integers 2 and -2 respectively. Thus the quotient set A/S can be identified with the set of integers. Addition and multiplication can be defined on A/S in terms of addition and multiplication on N.
   (d) The set A/S can be described geometrically as the set of circles centre the origin.
   (e) The members of A/S are sets of polynomials, of the form $\{p(x) + c \mid c \in R\}.$

2. (a) To prove that $T \circ S$ an equivalence relation implies $T \circ S = S \circ T$, use symmetry of S, T, $T \circ S$. To prove that $T \circ S = S \circ T$ implies $T \circ S$ is reflexive, symmetric and transitive, in each case use the corresponding properties of S and T.
   (b) $T \circ S$ is the relation of congruence modulo 2.

Exercises 3.5

2. The relations of (c) and (g) are total order relations.

3. Yes.

4. Suppose that c_1 and c_2 both satisfy the conditions for

l.u.b. $\{a, b\}$. Then $(c_1, c_2) \in S$ and $(c_2, c_1) \in S$ so that $c_1 = c_2$. Thus least upper bounds are unique.

(a) The set is a lattice with l.u.b. $\{A, B\} = A \cap B$ and g.l.b. $\{A, B\} = A \cup B$.

(b) The set is not a lattice since if $\arg z_1 \neq \arg z_2$ then $\forall z$, $(z_1, z) \notin S$ or $(z_2, z) \notin S$.

(c) The set is a lattice with l.u.b. $\{m, n\} = \max\{m, n\}$ and g.l.b. $\{m, n\} = \min\{m, n\}$.

(d) The set is a lattice, and is in fact totally ordered. An ordering constructed in this way is called a lexicographic ordering.

(e) The set is a lattice.

(f) The set is not a lattice since $\forall c \in A$, $(3, c) \notin S$ or $(10, c) \notin S$.

(g) The set is a lattice. Indeed all totally ordered sets are lattices.

5. Let θ be a fixed argument. Then $(z_0, z_0 e^{i\theta}) \in S$, and $z_0 \in [z]$ if and only if $z_0 e^{i\theta} \in [z]$. Thus $([z], [w]) \in T^*$ if and only if there are elements z_0, w_0 such that $z_0 e^{i\theta} \in [z]$, $w_0 e^{i\theta} \in [w]$ and $(z_0 e^{i\theta}, w_0 e^{i\theta}) \in T$. Each of the properties of reflexivity, antisymmetry and transitivity for T^* now follows from the corresponding property for T.

Exercises 3.7

1. Functions (b), (e) are injective. Functions (b), (d), (e), (f), (g), (h) are surjective.

4. The converse is not true. The following function is a counter-example.

$$f: x \rightsquigarrow \begin{cases} x, x \text{ rational,} \\ -x, x \text{ irrational.} \end{cases}$$

5. $T = \{(f, g) \mid \forall x \in \mathcal{D}, f(x) \leqslant g(x)\}$ is a partial order relation on A. With this relation A forms a lattice, with

$$\text{l.u.b.} \{f, g\} = \{(x, y) \mid x \in \mathcal{D} \text{ and } y = \max\{f(x), g(x)\}\}$$

$$\text{g.l.b.} \{f, g\} = \{(x, y) \mid x \in \mathcal{D} \text{ and } y = \min\{f(x), g(x)\}\}.$$

6. $231 = 3 \times 7 \times 11$ and so the set of factors of 231 has the same diagram as that of the set of factors of $30 = 2 \times 3 \times 5$. If A is a

set with a partial order relation S, and B is a set with a partial order relation T, then a bijection $f \colon A \rightsquigarrow B$ is said to preserve the partial ordering if and only if

$$\forall x, y \in A, (x, y) \in S \Rightarrow (f(x), f(y)) \in T.$$

The set of factors of a product of four prime factors, e.g. $2 \times 3 \times 5 \times 7$ will have the same diagram as $\{a, b, c, d\}$. Since $12 = 2 \times 2 \times 3$ and there is nothing in set theory corresponding to this repetition of the prime factor 2, no collection of subsets ordered by inclusion will have a branching diagram the same as that for the set of factors of 12. This conclusion can also be arrived at by counting the number of factors of 12, and using the result of exercise 7 of §1.4.

Exercises 3.8

1. Let the sets be $A_1, A_2, A_3, \ldots$, where $\forall i, A_i = \{a_{i1}, a_{i2}, a_{i3}, \ldots\}$. Repeat the enumeration of the rationals, replacing i/j by a_{ij}.

2. $f \colon A \rightsquigarrow B$, $f \colon x \rightsquigarrow c + (x-a)(d-c)/(b-a)$ is a bijection.

3. The sets $A \cap Q$ and $B \cap Q$ are denumerable and so are equivalent. The sets $A \backslash Q$ and $B \backslash Q$ are equal. Since $(A \cap Q) \cap (A \backslash Q) = \varnothing$, the sets $A = (A \cap Q) \cup (A \backslash Q)$ and B are equivalent.

4. If $A \backslash B$ and B are denumerable then so is A.

5. The set of lines joining pairs of points is finite, and so the set of gradients of these lines is finite. However, the set of gradients of all lines in the plane in non-denumerable. Thus there is a gradient θ such that no line with gradient θ contains more than one of the points. There are two million distinct lines of gradient θ which pass through points of the set, and these lines have a natural left to right ordering. There is a middle pair of such lines with respect to this ordering, and any line of gradient θ between this pair has the desired property.

7. The argument is not valid for $n = 2$.

8. $a_n = n + (-1)^n$.

9. (a), (b) and (e) can be proved to be true by induction. For (c), although $p_n \Rightarrow p_{n+1}$ is true, p_1 is false. For (d), $p_n \Rightarrow p_{n+1}$ is false, although p_n is true for $n = 1, 2, \ldots, 40$. Clearly p_{41} is false.

Exercises 4.2

1. (a) No. (b) Yes. (c) Yes. (d) Yes. (e) No. (f) No. (g) No. (h) Yes. (i) No.
2. Commutative: $f_1, f_2, f_4, f_6, f_7, f_9, f_{12}, f_{13}$.
   Associative: $f_1, f_2, f_4, f_5, f_6, f_7, f_9, f_{12}$.
   Identity elements: $\varnothing$ for f_1, U for f_2, $\varnothing$ for f_4, i_A for f_5, 0 for f_7, 1 for f_9, b for f_{14}.
   Inverse elements: every set is inverse to itself under f_4; every bijection has an inverse under f_5; every real number x has inverse $-x$ under f_7; every non-zero rational number x has inverse $1/x$ under f_9; every element has an inverse under f_{14}.
3. $\{((a_i, a_j), a_k) \mid k = (i+j) \bmod 4\}$. The operation is commutative, associative, has identity a_0 and each element a_i has inverse a_j where $i+j = 0 \bmod 4$.
4. If e_1 and e_2 are both identity elements for a binary operation f on X, then $f(e_1, e_2) = e_1$ and $f(e_1, e_2) = e_2$. Hence $e_1 = e_2$.
5. The binary operation f_{14} has identity element b, and both c and d are inverse to c.
6. f_1 is distributive over f_2 and f_2 is distributive over f_1. No other pairs are distributive. Note that f_9 is not distributive over f_6 because these operations have different domains.

Exercises 4.3

1. $(\mathscr{P}(U), f_4)$ and (R, f_7) are groups.
2. $(H_1 \cup H_2, *)$ need not be a subgroup.

Exercises 4.4

3. Propositions (i), (ii) are true. The conditions imply immediately that the points are distinct except possibly for q and s. It is is necessary to prove that $q \neq s$. Application of definition 4.8 (iv) together with the conditions leads to the inequalities

$$\rho(p,q) + \rho(q,s) \leqslant \rho(p,q) + \rho(q,r) + \rho(r,s) = \rho(p,s) \leqslant \rho(p,q) + \rho(q,s)$$

from which the result now follows.
Proposition (iii) is false. A counterexample is to be found in two

pairs of diametrically opposite points on a circle with shortest arc length as the distance function.

4. In Cauchy's and Minkowski's inequalities, equality occurs if and only if for some $k \neq 0$, $ku_i = v_i$ for all $i = 1, 2, \ldots, n$, i.e. if and only if the u_i are proportional to the v_i.

5. Properties (i), (ii), (iii) are trivial. To prove (iv), note that the function $x/(1+x)$ decreases as x decreases through positive values and so

$$\begin{aligned} \sigma(x, y) + \sigma(y, z) &= \frac{\rho(x, y) + \rho(y, z) + 2\rho(x, y)\rho(y, z)}{1 + \rho(x, y) + \rho(y, z) + \rho(x, y)\rho(y, z)} \\ &\geqslant \frac{\rho(x, y) + \rho(y, z) + \rho(x, y)\rho(y, z)}{1 + \rho(x, y) + \rho(y, z) + \rho(x, y)\rho(y, z)} \\ &\geqslant \frac{\rho(x, z)}{1 + \rho(x, z)} = \sigma(x, z). \end{aligned}$$

7. Properties (i), (ii), (iii) are trivial. Property (iv) is proved by applying Minkowski's inequality to the appropriate expression.

Exercises 4.5

2. (a) $A \cap B = [(A \cup B) \backslash (A \backslash B)] \backslash (B \backslash A) \in \mathscr{A}$.
   (b) $A \triangle B = (A \backslash B) \cup (B \backslash A) \in \mathscr{A}$.
   (c) Proved by induction.

3. $\mathscr{A} \cap \mathscr{B}$ and $\mathscr{P}(U)$ both satisfy (iii). $\mathscr{A} \cup \mathscr{B}$ and $\mathscr{A} \backslash \mathscr{B}$ need not satisfy (iii), as for example when $A \neq \varnothing$, $B \neq \varnothing$, $A \neq B$ and $\mathscr{A} = \{\varnothing, A, A^c, U\}$ and $\mathscr{B} = \{\varnothing, B, B^c, U\}$.

4. Apply Proposition 4.17 to $(A \cup B) \cup C$, and then the distributive law together with further applications of Proposition 4.17.

6. 100.

Exercises 5.2

1. The following models are possible.

   (a) $S = \{1, 2, 3, 4, 5, 6\}$, $P(\{i\}) = \dfrac{6-i}{21}$.

(e) $S = \{n \mid 1 \leqslant n \leqslant 1b\}$,
   $P(\{1\}) = P(\{5\}) = P(\{9\}) = P(\{10\}) = \frac{1}{49}$,
   $P(\{6\}) = \frac{7}{49}, P(\{3\}) = P(\{8\}) = \frac{4}{49}$,
   $P(\{2\}) = P(\{4\}) = \frac{15}{49}, P(\{7\}) = 0$.
(f) $S = \{Y, N\}, P(\{Y\}) = 0{\cdot}48, P(\{N\}) = 0{\cdot}52$.

2. (b) Prove by induction using Proposition 4.17.

(d) $P\left(S \setminus \bigcup_{i=1}^{n} A_i \right) = 1 - \sum_{i=1}^{n} P(A_i) + \sum_{i \neq j} P(A_i \cap A_j)$
   $- \sum_{i \neq j \neq k} P(A_i \cap A_j \cap A_k) + \ldots + (-1)^n P(A_1 \cap A_2 \cap \ldots \cap A_n)$.

Exercises 5.3

2. (a) $P(2 \mid \{2, 4, 6\}) = \frac{5}{8}$.   (e) $\frac{3}{11}$.
   (f) The model does not enable one to calculate such a probability.

3. (a) $\frac{5}{8}$.   (b) $\frac{3}{5}$.   (c) $\frac{9}{13}$.

5. $P(4 \mid d) \simeq 0{\cdot}68$.

INDEX